# Fractional Calculus

## An Introduction For Physicists

# Fractional Calculus

## An Introduction for Physicists

Richard Herrmann

GigaHedron, Germany

NEW JERSEY • LONDON • SINGAPORE • BEIJING • SHANGHAI • HONG KONG • TAIPEI • CHENNAI

*Published by*

World Scientific Publishing Co. Pte. Ltd.
5 Toh Tuck Link, Singapore 596224
*USA office:* 27 Warren Street, Suite 401-402, Hackensack, NJ 07601
*UK office:* 57 Shelton Street, Covent Garden, London WC2H 9HE

**British Library Cataloguing-in-Publication Data**
A catalogue record for this book is available from the British Library.

**FRACTIONAL CALCULUS**
**An Introduction for Physicists**



ISBN-13 978-981-4340-24-3
ISBN-10 981-4340-24-3

Printed in Singapore by World Scientific Printers.

# Foreword

Theoretical physics has evolved into a multi-faceted science. The sheer amount of knowledge accumulated over centuries forces a lecturer to focus on the mere presentation of derived results. The presentation of strategies or the long path to find appropriate tools and methods often has been neglected.

As a consequence, from a student's point of view, theoretical physics seems to be a construct of axiomatic completeness, where apparently is no space left for speculations and new approaches. The limitations of presented concepts and strategies are not obvious until these tools are applied to new problems.

Especially the creative process of research, which evolves beyond the state of mere reception of already known interrelations is a particular ambition of most students.

One prerequisite to reach that goal is a permanent readiness, to question even well-established results and check the validity of common statements time and time again. Already Roger Bacon in his *opus majus* pointed out, that the main causes of error are the belief in false authority, the force of habit, the ignorance of others and pretended knowledge [Bacon(1267)].

New concepts and new methods depend on each other. The unified theory of electromagnetism by James Clerk Maxwell was motivated by Faraday's experiments. The definite formulation was then realized in terms of a new theory of partial differential equations. Einstein's general relativity is based on experiments on the constancy of the speed of light; the elegant presentation was made possible with Riemann's tensor calculus.

This book too takes up an idea which is simple at first: due to the fact that differential equations play such a central role in physics since the times of Newton and since the first and second derivatives denote such fundamental properties as velocity and acceleration, does it make sense, to investigate the physical meaning of a half, a $\pi$-th and an imaginary derivative?

This interesting question has been raised since the days of Leibniz and has been discussed by all mathematicians of their times. But for centuries no practical applications could be imagined.

During the last years fractional calculus has developed rapidly. This process is still going on, but we can already recognize, that within the framework of fractional calculus new concepts and strategies emerge, which make it possible, to obtain new challenging insights and surprising correlations between different branches of physics.

This is the basic purpose of this book: to present a concise introduction to the basic methods and strategies in fractional calculus and to enable the reader to catch up with the state of the art on this field and to participate and contribute in the development of this exciting research area.

In contrast to other monographs on this subject, which mainly deal with the mathematical foundations of fractional calculus, this book is devoted to the application of fractional calculus on physical problems. The fractional concept is applied to subjects in classical mechanics, group theory, quantum mechanics, nuclear physics, hadron spectroscopy up to quantum field theory and will surprise the reader with new intriguing insights.

This book provides a skillful insight into a vividly growing research area and opens the reader's mind to a world yet completely unexplored. It encourages the reader to participate in the exciting quest for new horizons of knowledge.

Richard Herrmann
Dreieich, Germany
Summer, 2010

# Acknowledgments

As a young student, I had the opportunity to attend the Nobel Laureate Meeting at Lindau/Germany in 1982. Especially Paul Dirac's talk was impressive and highly inspiring and it affirmed my aspiration to focus my studies on theoretical physics.

I joined the Institute for Theoretical Physics in Frankfurt/Germany. Working on the fission properties of super-heavy elements, I suggested to linearise the collective Schrödinger equation used in nuclear collective models and to introduce a new collective degree of freedom, the collective spin. Within this project, my research activities are concentrated on the field theoretical implications, which follow from linearised wave equations. As a direct consequence the question of the physical interpretation of multi-factorized wave equations and of a fractional derivative came up.

I achieved a breakthrough in my research program in 2005: I found that a specific realization of the fractional derivative suffices to describe the properties of a fractional extension of the standard rotation group $SO(n)$. This was the key result to start an investigation of symmetries of fractional wave equations and the concept of a fractional group theory could be successfully realized. Up to now, this concept has led to a vast amount of intriguing and valuable results.

This success would not have been possible without the permanent enduring loving support of my family and friends.

I am grateful to all those who shared their knowledge with me and made suggestions reflected in this book. Special thanks go to Anke Friedrich and Günter Plunien for their encouragement and their valuable contributions.

I also want to emphasize, that fractional calculus is a worldwide activity. I benefited immensely from international support and communication. Discussions and correspondence with Eberhard Engel, Ervin Goldfain, Cesar

Ionescu, Swee Cheng Lim, Ahmad-Rami El-Nabulsi, Manuel Ortigueira and Volker Schneider were particularly useful.

Finally I want to express my gratitude to the publishing team of Word Scientific for their support.

# Contents

# Chapter 1

# Introduction

During recent years the interest of physicists in nonlocal field theories has been steadily increasing. The main reason for this development is the expectation that the use of these field theories will lead to a much more elegant and effective way of treating problems in particle and high-energy physics as it has been possible up to now with local field theories.

Nonlocal effects may occur in space and time. For example in the time domain the extension from a local to a nonlocal description becomes manifest as a memory effect, which roughly states that the actual behaviour of a given object is not only influenced by the actual state of the system but also by events, which happened in the past.

In a first approach this could be interpreted as an ability of the object to collect or memorize previous events, an idea going back to the physics of Aristotle and perpetuated until the scholastic era. But this concept already irritated Descartes [Descartes(1664)] and seems obsolete since the days of Newton.

In order to allow a better understanding of nonlocal effects, we consider as a simple example the motion of a classical particle in a diluted gas. On the left side of Fig. 1.1 we illustrate the free field case which is characterized by the absence of any boundaries. In that case, with a given collision rate $\psi(t)$, the dynamics of the system is determined by a local theory.

Let us now introduce some walls or boundaries, where the gas molecules bounce off.

This situation is schematically sketched on the right of Fig. 1.1. In that case, at the time $t - \tau$ a fixed rate of gas molecules is scattered, but in contrast to the free field case, at the time $t - \tau/2$ they collide with the boundaries and are reflected. At the time $t$ the dynamics of the system is then characterized by a source term which besides $\psi(t)$ contains an additional nonlocal term proportional to $\psi(t - \tau)$. This is a simple geometric

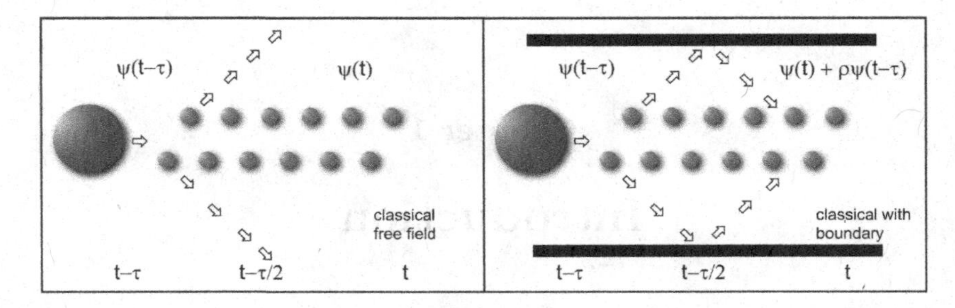


Fig. 1.1. A possible geometric interpretation of a memory effect. The boundary-free case shown on the left side is described by a local theory (characterized by a source term $\psi(t)$), while the same case including boundaries may be described by a nonlocal theory (characterized by a source term $\psi(t) + \rho\psi(t-\tau)$ with $\rho$ being a proportionality factor).

interpretation of a nonlocal theory. Neither the described particle nor the surrounding medium memorize previous events. There is no intelligence in the system. The boundaries generate a delayed reaction of the medium, which results in a memory effect.

In a more sophisticated scenario, we may assert to every point in space a complex reflection coefficient and are led to Huygens' principle which states that each point of an advancing wavefront is in fact the source of a new set of waves.

In that sense a nonlocal theory may be interpreted as a construct which allows a smooth transition between a local (Newtonian) and a full quantum theory of motion.

A particular subgroup of nonlocal field theories plays an increasingly important role, may be described with operators of fractional nature and is specified within the framework of fractional calculus.

From a historical point of view fractional calculus may be described as an extension of the concept of a derivative operator from integer order $n$ to arbitrary order $\alpha$, where $\alpha$ is a real or complex value, or even more complicated a complex valued function $\alpha = \alpha(x, t)$:

$$\frac{d^n}{dx^n} \to \frac{d^\alpha}{dx^\alpha} \tag{1.1}$$

Despite the fact that this concept is being discussed since the days of Leibniz [Leibniz(1695)] and since then has occupied the great mathematicians of their times, no other research area has resisted a direct application for centuries. Abel's treatment of the tautochrone problem [Abel(1823)] from 1823 stood for a long time as a singular example for an application of fractional calculus.

Not until the works of Mandelbrot [Mandelbrot(1982)] on fractal geometry in the early 1980's was the interest of physicists attracted by this subject, and a first wave of publications in the area of fractional Brownian motion and anomalous diffusion processes was created. But these works lead only to possibly a handful of useful applications and did not produce results of far-reaching consequences [Kilbas(2003)].

The situation changed drastically by progress made in the area of fractional wave equations in recent years. Within this process, new questions in fundamental physics have been raised, that cannot be formulated adequately using traditional methods. Consequently a new research area has emerged, that allows for new insights and intriguing results using new methods and approaches.

The interest in fractional wave equations was amplified in 2000 with a publication by Raspini [Raspini(2000)]. He deduced an $SU(3)$ symmetric wave equation, which turned out to be of fractional nature. In contrast to this formal derivation a standard Yang–Mills theory is merely a recipe for coupling any phenomenologically deduced symmetry. Zavada [Zavada(2002)] has generalized Raspini's result: he demonstrated, that an $n$-fold factorization of the d'Alembert operator automatically leads to fractional wave equations with an inherent $SU(n)$ symmetry.

In 2002, Laskin [Laskin(2002)] on the basis of the Riesz definition [Riesz(1949)] of the fractional derivative presented a Schrödinger equation with fractional derivatives and gave a proof of Hermiticity and parity conservation of this equation.

In 2005, the Casimir operators and multiplets of the fractional extension of the standard rotation group $SO(n)$ were calculated algebraically [Herrmann(2005)]. A mass formula was derived, which successfully described the ground state masses of the charmonium spectrum. This may be interpreted as a first approach to investigate a fractional generalization of a standard Lie algebra, a first attempt to establish a fractional group theory and the first non-trivial application of fractional calculus in multidimensional space.

In 2006, Goldfain [Goldfain(2006)] demonstrated, that the low level fractional dynamics [Tarasov(2006)] on a flat Minkowski metric most probably describes similar phenomena as a field theory in curved Riemann space time. In addition, he proposed a successful mechanism to quantize fractional free fields. Lim [Lim(2006a)] proposed a quantization description for free fractional fields of Klein–Gordon-type and investigated the temperature dependence of those fields.

In 2007, we [Herrmann(2007)] applied the concept of local gauge invariance to fractional free fields and derived the exact interaction form in the first order of the coupling constant. The fractional analogue of the normal Zeeman-effect was calculated and as a first application a mass formula was presented, which gives the masses of the baryon spectrum an accuracy better than 1%. It has been demonstrated, that the concept of local gauge invariance determines the exact form of interaction, which in the lowest order coincides with the derived group chain for the fractional rotation group.

Since then the investigation of the fractional rotation group alone within the framework of fractional group theory has lead to a vast amount of interesting results, e.g. a theoretical foundation of magic numbers in atomic nuclei and metallic clusters.

It is the explicit intention of this book to give a concise introduction to this new vividly growing branch of fractional calculus.

This book is an invitation to the interested student and to the professional researcher as well. It presents a thorough introduction to the basics of fractional calculus and routes the reader up to the current state-of-the-art physical interpretation.

What makes this textbook unique is its application oriented approach. If you are looking merely for a mathematical foundation of fractional calculus we refer to the literature (e.g. [Oldham(1974), Samko(1993b), Miller(1993), Kiryakova(1994), Rubin(1996), Gorenflo(1997), Podlubny(1999), Hilfer(2000, 2008), Mainardi(2010)]). This book is explicitly devoted to the practical consequences of using fractional calculus.

The book may be divided into two parts. Chapters 2 to 8 give a step by step introduction to the techniques and methods derived in fractional calculus and their application to classical problems.

Chapters 9 to 19 are devoted to a concise introduction of fractional calculus in the area of quantum mechanics of multi-particle systems. The application of group theoretical methods will lead to new and unexpected results. The reader is directly led to the actual state of research.

All derived results are directly compared to experimental findings. As a consequence, the reader is guided on a solid basis and is encouraged to apply the fractional calculus approach in his research area, too.

It will be demonstrated, that the viewpoint of fractional calculus leads to new insights and surprising interrelations of classical fields of research that remain unconnected until now.

## Chapter 2

# Functions

*What is a number?*

*When the Pythagoreans believed that everything is a number they had in mind primarily integer numbers or ratios of integers [Burkert(1972)]. The discovery of additional number types, first ascribed to Hippasus of Metapontum, at that time must have been both, a deeply shocking as well as an extremely exciting experience which in the course of time helped to gain a deeper understanding of mathematical problems.*

*The increasing commercial success and volume of trade in the North Italian city states commencing on the reign of the Stauffian emperors promoted the progress of practical mathematics and gave a first application of the negative numbers in terms of a customer's debts [Fibonacci(1204)].*

*The general solutions of cubic and quartic equations forced an interpretation of imaginary numbers [Bombelli(1572)], which were finally accepted as useful quantities, when Gauss gave a general interpretation of the fundamental theorem of algebra [Gauß(1799)] not until 1831.*

*If we consider a number as a scalar object, we may proceed and introduce a vector, a matrix or a general tensor as objects with increasing level of complexity regarding their transformation properties. Therefore we have a well defined sequence of definitions of number with increasing complexity.*

*Furthermore we have nice examples for a successful extension of the scope of functions e.g. from integer to real values. Let us begin with two stories of success: we present the gamma and Mittag-Leffler functions, which turn out to be well established extensions of the factorial and the exponential function. These functions play an important role for practical applications of the fractional calculus.*

## 2.1 Gamma function

The gamma function is the example par excellence for a reasonable extension of the scope of a function from integer to real up to imaginary numbers.

For natural numbers the factorial $n!$ is defined by

$$n! = 1 \times 2 \times 3 \times ... \times n = \prod_{j=1}^{n} j \tag{2.1}$$

with the following essential properties

$$1! = 1 \tag{2.2}$$

$$n! = n(n-1)! \tag{2.3}$$

Now we define the gamma function on the set of complex numbers $z$

$$\Gamma(z) = \lim_{n\to\infty} = \frac{n!n^z}{z(z+1)(z+2)...(z+n)} \tag{2.4}$$

or alternatively

$$\frac{1}{\Gamma(z)} = ze^{\gamma z} \prod_{n=0}^{\infty} \left(1 + \frac{z}{n}\right) e^{-z/n} \tag{2.5}$$

where $\gamma = 0.57721...$ denotes the Euler constant.

The scope of both definitions extends on the field of complex numbers $z$.

In case $\mathrm{Re}(z) > 0$ the integral representation

$$\Gamma(z) = \int_0^{\infty} t^{z-1} e^{-t} dt \tag{2.6}$$

is valid.

From integration by parts it follows

$$\Gamma(1+z) = z\Gamma(z) \tag{2.7}$$

and by direct integration $\Gamma(1) = 1$. Therefore we associate:

$$\Gamma(1+n) = n! \tag{2.8}$$

Hence the gamma function may be considered a reasonable extension of the factorial.

Within the framework of fractional calculus, as we shall see later, the zeroes of the reciprocal value of the gamma function are of particular interest:

$$\frac{1}{\Gamma(z)} = 0 \qquad z = -n \quad n = 0, 1, 2, 3... \tag{2.9}$$

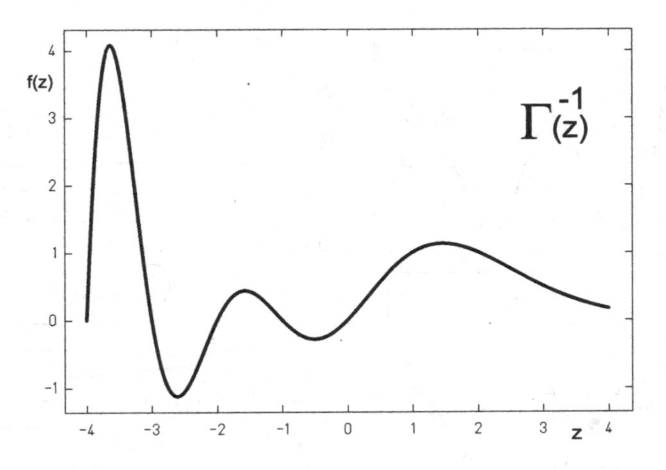


Fig. 2.1. The inverse gamma function $f(z) = \frac{1}{\Gamma(z)}$.

Furthermore we may extend the definition of the binomial coefficients:

$$\binom{z}{v} = \frac{z!}{v!(z-v)!} = \frac{\Gamma(1+z)}{\Gamma(1+v)\Gamma(1+z-v)} \tag{2.10}$$

In addition the reflection formula

$$\Gamma(z)\Gamma(1-z) = \frac{\pi}{\sin(\pi z)} \tag{2.11}$$

will be used.

## 2.2 Mittag-Leffler functions

Besides the gamma function Euler has brought to light an additional important function, the exponential:

$$e^z = \sum_{n=0}^{\infty} \frac{z^n}{n!} \tag{2.12}$$

According to our remarks made in the previous section we may replace the factorial by the gamma function:

$$e^z = \sum_{n=0}^{\infty} \frac{z^n}{\Gamma(1+n)} \tag{2.13}$$

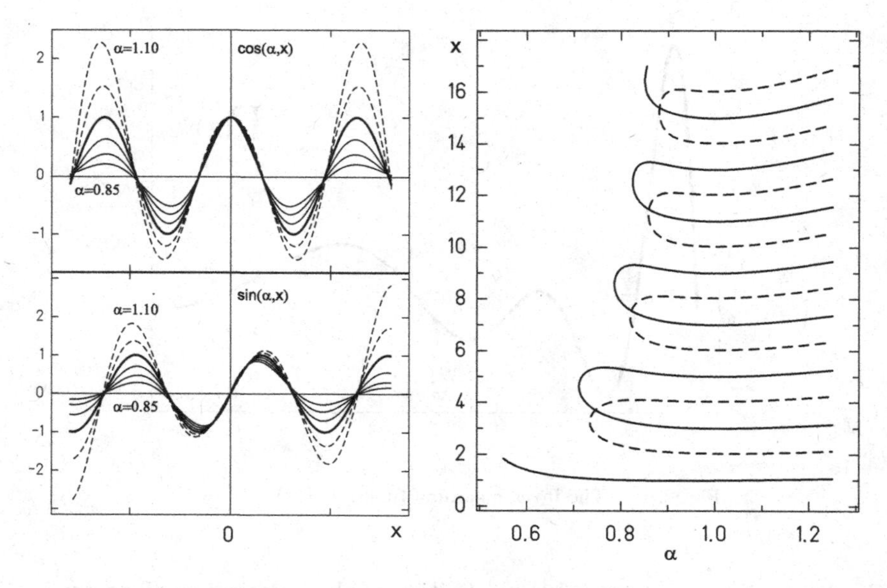


Fig. 2.2. Solutions and zeroes of the Caputo-wave equation are the fractional pendant of the trigonometric functions and special cases of the Mittag-Leffler function $\cos(\alpha, x) = E(2\alpha, -x^{2\alpha})$ and the generalized Mittag-Leffler function $\sin(\alpha, x) = x^{\alpha} E(2\alpha, 1+\alpha, -x^{2\alpha})$. The graph is given for different $\alpha$ near $\alpha = 1$. Units are given as multiples of $\pi/2$.

Without difficulty this definition may be extended, where one option is given by:

$$E(\alpha, z) = \sum_{n=0}^{\infty} \frac{z^n}{\Gamma(1 + n\alpha)} \tag{2.14}$$

where we have introduced an arbitrary real number $\alpha > 0$. The properties of this function $E(\alpha, z)$ were first investigated systematically in the year 1903 by Mittag-Leffler and consequently it is called Mittag-Leffler function [Mittag-Leffler(1903)]. In the area of fractional calculus this function is indeed of similar importance as is the exponential function in standard mathematics and is subject of actual research [Gorenflo(1996), Mainardi(2000), Hilfer(2006)]. An extensive collection of standard functions are special cases of the Mittag-Leffler function:

$$E(2, -z^2) = \cos(z), \qquad E(1/2, z^{\frac{1}{2}}) = e^z \left(1 + \mathrm{erf}(z^{\frac{1}{2}})\right) \tag{2.15}$$

where the error function $\mathrm{erf}(z)$ is given by

$$\mathrm{erf}(z) = \frac{2}{\sqrt{\pi}} \int_0^z e^{-t^2} dt \tag{2.16}$$

An extended version of the Mittag-Leffler function is given by

$$E(\alpha, \beta, z) = \sum_{n=0}^{\infty} \frac{z^n}{\Gamma(n\alpha + \beta)} \tag{2.17}$$

This function is named generalized Mittag-Leffler function [Wiman(1905), Seybold(2005), Hilfer(2006), Seybold(2008)]. It is one of the many reasonable extensions of the exponential function. A large number of simple functions are special cases of the generalized Mittag-Leffler function, for example:

$$E(1, 2, z) = \frac{e^z - 1}{z}, \qquad E(2, 2, z^2) = \frac{\sinh(z)}{z} \tag{2.18}$$

## 2.3 Hypergeometric functions

Hypergeometric functions contain many particular special functions as special cases and are used extensively within the framework of fractional calculus. From the definition

$$_pF_q(\{a_i\}; \{b_j\}; z) = \frac{\prod_{j=1}^{q} \Gamma(b_j)}{\prod_{i=1}^{p} \Gamma(a_i)} \sum_{n=0}^{\infty} \frac{\prod_{i=1}^{p} \Gamma(a_i + n)}{\prod_{j=1}^{q} \Gamma(b_j + n)} \frac{z^n}{n!} \tag{2.19}$$

follows for the derivative

$$\frac{d}{dz}\, {}_pF_q(\{a_i\}; \{b_j\}; z) = \frac{\prod_{i=1}^{p} \Gamma(a_i)}{\prod_{j=1}^{q} \Gamma(b_j)}\, {}_pF_q(\{a_i + 1\}; \{b_j + 1\}; z) \tag{2.20}$$

Beyond many other special cases hypergeometric functions include the Gauss' s hypergeometric series ${}_2F_1(a, b; c; z)$ and the confluent hypergeometric series ${}_1F_1(a; b; z)$ which is a solution of Kummer's differential equation.

## 2.4 Miscellaneous functions

Fractional calculus not only extends our view on physical processes. It also breaks into regions of mathematics, which in general are neglected in physics. Since we will derive in the course of this book some exotic results, an alphabetic listing of functions and their notation is given here:

**complementary error function [Wolfram(2010)]**

$$\operatorname{erfc}(z) = 1 - \operatorname{erf}(z) \tag{2.21}$$

**cosine integral [Abramowitz(1965)]**

$$\mathrm{Ci}(z) = -\int_z^\infty \frac{\cos(t)}{t}dt = \gamma + \ln(z) + \int_0^z \frac{\cos(t)-1}{t}dt \tag{2.22}$$

**digamma function [Abramowitz(1965)]**

$$\Psi(z) = \frac{d}{dz}\ln(\Gamma(z)) \tag{2.23}$$

**Euler $\gamma$ constant [Wolfram(2010)]**

$$\gamma = 0.577215... \tag{2.24}$$

**exponential integral Ei [Spanier(2008)]**

$$\mathrm{Ei}(z) = -\int_{-z}^\infty \frac{e^{-t}}{t}dt \tag{2.25}$$

**exponential integral E$_n$ [Wolfram(2010)]**

$$\mathrm{E}_n(z) = -\int_1^\infty \frac{e^{-zt}}{t^n}dt \tag{2.26}$$

**harmonic number [Wolfram(2010)]**

$$H_z = \gamma + \Psi(z+1) = z_3F_2(1,1,1-z,2,2,1) \tag{2.27}$$

**Hurwitz function [Hurwitz(1882)]**

$$\zeta(s,z) = \sum_{k=0}^\infty (k+z)^{-s} \tag{2.28}$$

**incomplete gamma function [Wolfram(2010)]**

$$\Gamma(a,z) = \int_z^\infty t^{a-1}e^{-t}dt \tag{2.29}$$

**Meijer G-function [Meijer(1941), Prudnikov(1990)]**

$$G^{m,n}_{p,q}\left(z\middle|\begin{matrix}a_1,...,a_p\\ b_1,...,b_q\end{matrix}\right) = \tag{2.30}$$

$$\frac{1}{2\pi i}\int_{\gamma L}\frac{\left(\prod_{j=1}^m \Gamma(b_j+s)\right)\left(\prod_{j=1}^n \Gamma(1-a_j-s)\right)}{\left(\prod_{j=n+1}^p \Gamma(a_j+s)\right)\left(\prod_{j=m+1}^q \Gamma(1-b_j-s)\right)}z^{-s}ds$$

where the contour $\gamma L$ lies between the poles of $\Gamma(1-a_j-s)$ and $\Gamma(b_j+s)$.

**polygamma function [Abramowitz(1965)]**

$$\Psi^{(n)}(z) = \frac{d^n}{dz^n}\Psi(z) = \frac{d^{n+1}}{dz^{n+1}}\ln(\Gamma(z)) \tag{2.31}$$

**sine integral [Abramowitz(1965)]**

$$\mathrm{Si}(z) = \int_0^z \frac{\sin(t)}{t}dt \tag{2.32}$$

# Chapter 3

# The Fractional Derivative

*The historical development of fractional calculus begins with a step by step introduction of the fractional derivative for special function classes. We follow this path and present these functions and the corresponding derivatives.*

*In addition we will present alternative concepts to introduce a fractional derivative definition, e.g. a series expansion in terms of the standard derivative. Furthermore we introduce the fractional extension of the Leibniz product rule and we will demonstrate, that an application of this rule will extend the range of function classes a fractional derivative may be defined for.*

*The strategies we present for an appropriate extension of the standard derivative definition may be applied uniquely in areas of mathematical sciences, where similar concepts are valid.*

*As one example we mention the important field of orthogonal polynomials, which is a vivid and actual research area.*

## 3.1 Basics

If we consider an application of differential or integral calculus simply as a mapping from a given function set $f$ onto another set $g$ e.g.

$$g(x) = \frac{d}{dx} f(x) \tag{3.1}$$

then in general from this relation we cannot deduce any valid information on a possible similarity of a function and its derivative.

Therefore it is surprising and remarkable that for particular function classes we observe a very simple relationship in respect of their derivatives.

For the exponential function we obtain

$$\frac{d}{dx}e^{kx} = ke^{kx} \tag{3.2}$$

for trigonometric functions we have

$$\begin{aligned}
\frac{d}{dx}\sin(kx) &= k\cos(kx) = k\sin\left(kx + \frac{\pi}{2}\right) \\
\frac{d}{dx}\left(k\cos(kx)\right) &= \tfrac{d^2}{dx^2}\sin(kx) = -k^2\sin(kx) = k^2\sin(kx+\pi) \\
\frac{d}{dx}\left(-k^2\sin(kx)\right) &= \tfrac{d^3}{dx^3}\sin(kx) = -k^3\cos(kx) = k^3\sin\left(kx + \frac{3\pi}{2}\right) \\
\frac{d}{dx}\left(-k^3\cos(kx)\right) &= \tfrac{d^4}{dx^4}\sin(kx) = k^4\sin(kx) = k^4\sin(kx+2\pi)
\end{aligned} \tag{3.3}$$

or for powers

$$\frac{d}{dx}x^k = kx^{k-1} \tag{3.4}$$

We deduce a simple rule which we can easily write in a generalized form valid for all $n \in \mathbb{N}$:

$$\frac{d^n}{dx^n}e^{kx} = k^n e^{kx} \tag{3.5}$$

$$\frac{d^n}{dx^n}\sin(kx) = k^n \sin\left(kx + \frac{\pi}{2}n\right) \tag{3.6}$$

$$\frac{d^n}{dx^n}x^k = \frac{k!}{(k-n)!}x^{k-n} \tag{3.7}$$

For arbitrary order $n$ apparently a kind of self similarity emerges, e.g. all derivatives of the exponential lead to exponentials, all derivatives of trigonometric functions lead to trigonometric functions. Since the derivative is given in this closed form it is straightforward to extend this rule from integer derivative coefficients $n \in \mathbb{N}$ to real and even imaginary coefficients $\alpha$ and postulate a fractional derivative as:

$$\frac{d^\alpha}{dx^\alpha}e^{kx} = k^\alpha e^{kx}, \qquad k \geq 0 \tag{3.8}$$

$$\frac{d^\alpha}{dx^\alpha}\sin(kx) = k^\alpha \sin\left(kx + \frac{\pi}{2}\alpha\right), \qquad k \geq 0 \tag{3.9}$$

$$\frac{d^\alpha}{dx^\alpha}x^k = \frac{\Gamma(1+k)}{\Gamma(1+k-\alpha)}x^{k-\alpha}, \qquad x \geq 0, k \neq -1, -2, -3, \ldots \tag{3.10}$$

We restrict to $k \geq 0$ and $x \geq 0$ respectively to ensure the uniqueness of the fractional derivative definition. This intuitive approach follows the historical path along which the fractional calculus has evolved. The definition of the fractional derivative of the exponential function was given by Liouville [Liouville(1832)]. The corresponding derivative of trigonometric functions has first been proposed by Fourier [Fourier(1822)]. The fractional derivative for powers was studied systematically by Riemann [Riemann(1847)]; first attempts to solve this problem were already made by Leibniz and Euler [Euler(1738)].

The fractional derivative of a constant function is given according to Riemann's definition (3.10) by:

$$\frac{d^\alpha}{dx^\alpha}\,\text{const} = \frac{d^\alpha}{dx^\alpha}x^0 = \frac{1}{\Gamma(1-\alpha)}x^{-\alpha} \tag{3.11}$$

which is an at first unexpected behaviour. Hence as an additional postulate Caputo has introduced

$$\frac{d^\alpha}{dx^\alpha}\,\text{const} = 0 \tag{3.12}$$

which allows for a more familiar definition of e.g. series expansions, as will be demonstrated in the following.

As a consequence we obtain four different definitions of a fractional derivative. However they share common aspects. On one hand all definitions fulfil a correspondence principle:

$$\lim_{\alpha\to n}\frac{d^\alpha}{dx^\alpha}f(x) = \frac{d^n}{dx^n}f(x), \quad n = 0, 1, 2, ... \tag{3.13}$$

In addition the following rules are valid:

$$\frac{d^\alpha}{dx^\alpha}cf(x) = c\frac{d^\alpha}{dx^\alpha}f(x) \tag{3.14}$$

$$\frac{d^\alpha}{dx^\alpha}(f(x)+g(x)) = \frac{d^\alpha}{dx^\alpha}f(x) + \frac{d^\alpha}{dx^\alpha}g(x) \tag{3.15}$$

These are very important properties, since we can define, in analogy to the well known Taylor series expansion series on these four different function classes and specify the corresponding fractional derivatives:

For the fractional derivative according to Liouville:

$$f(x) = \sum_{k=0}^{\infty} a_k e^{kx} \tag{3.16}$$

$$\frac{d^\alpha}{dx^\alpha}f(x) = \sum_{k=0}^{\infty} a_k k^\alpha e^{kx} \tag{3.17}$$

the fractional derivative according to Fourier:

$$f(x) = a_0 + \sum_{k=1}^{\infty} a_k \sin(kx) + \sum_{k=1}^{\infty} b_k \cos(kx) \tag{3.18}$$

$$\frac{d^\alpha}{dx^\alpha}f(x) = \sum_{k=1}^{\infty} a_k k^\alpha \sin\left(kx + \frac{\pi}{2}\alpha\right) + \sum_{k=1}^{\infty} a_k k^\alpha \cos\left(kx + \frac{\pi}{2}\alpha\right) \tag{3.19}$$

the fractional derivative according to Riemann:

$$f(x) = x^{\alpha-1} \sum_{k=0}^{\infty} a_k x^{k\alpha} \tag{3.20}$$

$$\frac{d^\alpha}{dx^\alpha} f(x) = x^{\alpha-1} \sum_{k=0}^{\infty} a_{k+1} \frac{\Gamma((k+2)\alpha)}{\Gamma((k+1)\alpha)} x^{k\alpha} \tag{3.21}$$

and for the fractional derivative according to Caputo:

$$f(x) = \sum_{k=0}^{\infty} a_k x^{k\alpha} \tag{3.22}$$

$$\frac{d^\alpha}{dx^\alpha} f(x) = \sum_{k=0}^{\infty} a_{k+1} \frac{\Gamma(1+(k+1)\alpha)}{\Gamma(1+k\alpha)} x^{k\alpha} \tag{3.23}$$

A given analytic function may therefore be expanded in a series and the fractional derivative may then be calculated.

With this result we can prove whether the four different definitions of a fractional derivative are equivalent. Do we obtain similar results for the fractional derivative for a given function?

We choose the exponential function as an example and calculate the fractional derivative according to Caputo:

$$\frac{d^\alpha}{dx^\alpha} e^x = \frac{d^\alpha}{dx^\alpha} \sum_{n=0}^{\infty} \frac{x^n}{n!} \tag{3.24}$$

$$= \sum_{n=1}^{\infty} \frac{x^{n-\alpha}}{\Gamma(1+n-\alpha)} \tag{3.25}$$

$$= x^{1-\alpha} E(1, 2-\alpha, x) \tag{3.26}$$

In contrast to Liouville's definition the fractional derivative according to Caputo's definition is obviously not an exponential function any more but a generalized Mittag-Leffler function (2.17).

We obtain the disturbing result, that the proposed definitions of a fractional derivative are not interchangeable, but will lead to different results if applied to the same function. Up to now there is no ultimate definition of a fractional derivative, even worse over time some additional definitions have been proposed which we will discuss extensively in the following chapters.

For practical applications of fractional calculus this leads to a situation, that e.g. the solutions of a fractional differential equation are unique, but will show different behaviour depending on the specific definition of the fractional derivative used. From a mathematical point of view all these definitions are equally well-founded. But in contrast to mathematicians physicists are in an advantageous position. Since they intend to describe natural phenomena they have the opportunity to compare theoretical results with experimental data.

In the following chapters we will therefore lay particular emphasis on the application of derived results and we will present evidence for the validity of one or the other formal definition of a fractional derivative. Nevertheless we have to admit that until now there is no unique candidate which explains all experimental data. Instead, depending on the specific problem, one of the fractional derivatives will lead to better agreement with the experiment.

A substantial progress in the theory of fractional calculus without doubt would be achieved if an extended formulation of a fractional derivative would be proposed, which was valid for more or less different function classes simultaneously.

A step in this direction is based on the idea, to deduce a series expansion of the fractional derivative in terms of the standard derivative. For that purpose we start with the well known series expansion:

$$(a+b)^n = \sum_{k=0}^{n} \binom{n}{k} a^{n-k} b^k = \sum_{k=0}^{n} \binom{n}{k} a^n b^{n-k} \tag{3.27}$$

which is valid for $n \in \mathbb{N}$. Using the fractional extension of the Binomial coefficient (2.10) we propose the following generalization which is valid for arbitrary powers $\alpha$:

$$(a+b)^\alpha = \sum_{k=0}^{\infty} \binom{\alpha}{k} a^{\alpha-k} b^k = \sum_{k=0}^{\infty} \binom{\alpha}{k} a^k b^{\alpha-k} \tag{3.28}$$

Hence we may present a definition of a fractional derivative as a series expansion:

$$\left(\frac{d}{dx}+\omega\right)^\alpha = \sum_{k=0}^{\infty} \binom{\alpha}{k} \omega^{\alpha-k} \frac{d^k}{dx^k} \tag{3.29}$$

where $\omega$ is an arbitrary number.

At first sight this result seems nothing else but one more definition of a fractional derivative. But at least for the fractional derivative according to Liouville the relation

$$\left(\frac{d}{dx}+\omega\right)^\alpha f(x) = e^{-\omega x} \frac{d^\alpha}{dx^\alpha} e^{\omega x} f(x) \tag{3.30}$$

holds.

We therefore obtain the encouraging result, that both definitions are strongly related.

In addition, the series expansion of the fractional derivative gives a first impression of the non-local properties of this operation. Since the sum of derivatives is infinite, we obtain an infinite series of derivative values which is equivalent to determine the full Taylor series of a given function. Depending on the convergence radius of this Taylor series this implies, that a knowledge about the exact behaviour of the function considered is necessary in order to obtain its fractional derivative.

In the following chapters we will demonstrate, that a large number of problems may already be solved using the proposed definitions of a fractional derivative (3.8)–(3.12).

## 3.2 The fractional Leibniz product rule

In the previous section we have presented a simple approach to fractional derivatives. But we want to state clearly that well-established techniques used in ordinary differential calculus may not be transferred thoughtlessly and without proof to the field of fractional calculus.

A typical example is the Leibniz rule in its familiar form

$$\frac{d}{dx}(\psi\,\chi) = \left(\frac{d}{dx}\psi\right)\chi + \psi\left(\frac{d}{dx}\chi\right) \tag{3.31}$$

It could be carelessly assumed, that in the case of fractional calculus the relation

$$\frac{d^\alpha}{dx^\alpha}(\psi\,\chi) = \left(\frac{d^\alpha}{dx^\alpha}\psi\right)\chi + \psi\left(\frac{d^\alpha}{dx^\alpha}\chi\right) \tag{3.32}$$

will hold.

A simple counter-example using e.g. the Liouville definition of a fractional derivative will do:

$$\frac{d^\alpha}{dx^\alpha}(e^{k_1 x}e^{k_2 x}) = \frac{d^\alpha}{dx^\alpha}e^{(k_1+k_2)x} \tag{3.33}$$

$$= (k_1+k_2)^\alpha e^{(k_1+k_2)x} \tag{3.34}$$

On the other hand with (3.32) we obtain:

$$\frac{d^\alpha}{dx^\alpha}(e^{k_1 x}e^{k_2 x}) = (k_1^\alpha + k_2^\alpha)(e^{k_1 x}e^{k_2 x}) \tag{3.35}$$

Therefore the Leibniz product rule according to (3.32) is not valid for fractional derivatives.

In order to derive an extension of this product rule which is valid for fractional derivatives too, let us first note the rule for higher derivatives:

$$\begin{aligned}
\frac{d}{dx}(\psi\,\chi) &= \left(\frac{d}{dx}\psi\right)\chi + \psi\left(\frac{d}{dx}\chi\right) \\
\frac{d^2}{dx^2}(\psi\,\chi) &= \left(\frac{d^2}{dx^2}\psi\right)\chi + 2\left(\frac{d}{dx}\psi\right)\left(\frac{d}{dx}\chi\right) + \psi\left(\frac{d^2}{dx^2}\chi\right) \\
\frac{d^3}{dx^3}(\psi\,\chi) &= \left(\frac{d^3}{dx^3}\psi\right)\chi + 3\left(\frac{d^2}{dx^2}\psi\right)\left(\frac{d}{dx}\chi\right) + 3\left(\frac{d}{dx}\psi\right)\left(\frac{d^2}{dx^2}\chi\right) \\
&\quad + \psi\left(\frac{d^3}{dx^3}\chi\right)
\end{aligned} \tag{3.36}$$

using the abbreviation

$$\partial_x^\alpha = \frac{d^\alpha}{dx^\alpha} \tag{3.37}$$

we obtain for arbitrary integers:

$$\partial_x^n(\psi\,\chi) = \sum_{j=0}^{n} \binom{n}{j} (\partial_x^{n-j}\psi)(\partial_x^j\chi) \qquad n \in \mathbb{N} \tag{3.38}$$

According to (3.28) this expression may be extended to arbitrary $\alpha$:

$$\partial_x^\alpha(\psi\,\chi) = \sum_{j=0}^{\infty} \binom{\alpha}{j} (\partial_x^{\alpha-j}\psi)(\partial_x^j\chi) \tag{3.39}$$

Indeed this form of the Leibniz product rule is valid for fractional derivatives too.

Of course there are many other known rules which are common in standard calculus e.g. the chain rule. As demonstrated, a direct transfer of any standard rule to fractional calculus is hazardous, instead in every case the validity of a rule used in fractional calculus must be checked carefully.

Therefore we have demonstrated, that a useful strategy for a satisfactory fractionalization scheme is given as a two step procedure:

- Derive a rule which is valid for all $n \in \mathbb{N}$.
- Replace $n$ by $\alpha \in \mathbb{C}$.

## 3.3 Discussion

### 3.3.1 *Orthogonal polynomials*

**Question:**
You have presented a procedure to generalize

- a) a function like factorial $n!$,
- b) the concept of a derivative or
- c) the Leibniz product rule

from integer values $n$ to arbitrary $\alpha$. There is another area in mathematical physics where integers play an important role in counting orthogonal polynomials. Will the proposed fractionalization procedure lead to sets of fractional orthogonal polynomials?

**Answer:**
The proposed fractionalization mechanism may indeed be applied to orthogonal polynomials leading to generalized polynomials of fractional order. We present an elementary example:

The Laguerre polynomials $L_n(x)$ are explicitly given by

$$L_n(x) = \sum_{m=0}^{n} (-1)^m \binom{n}{n-m} \frac{1}{m!} x^m \tag{3.40}$$

$$= {}_1F_1(-n; 1; x) \tag{3.41}$$

They obey the differential equation:

$$x \frac{d^2}{dx^2} L_n(x) + (1-x) \frac{d}{dx} L_n(x) + n L_n(x) = 0 \quad , n \in \mathbb{N} \tag{3.42}$$

and the orthogonality relation is given by

$$\int_0^\infty e^{-x} L_n(x) L_m(x) dx = \delta_{nm} \tag{3.43}$$

As a generalization from integer $n$ to arbitrary $\alpha$ we propose

$$L_\alpha(x) = \sum_{m=0}^{\infty} (-1)^m \binom{\alpha}{\alpha-m} \frac{1}{m!} x^m \tag{3.44}$$

$$= {}_1F_1(-\alpha; 1; x) \tag{3.45}$$

Indeed it can be proven, that the fractional Laguerre polynomials $L_\alpha(x)$ fulfil the differential equation

$$x \frac{d^2}{dx^2} L_\alpha(x) + (1-x) \frac{d}{dx} L_\alpha(x) + \alpha L_\alpha(x) = 0 \quad , \alpha \in \mathbb{C} \tag{3.46}$$

which is nothing else but a generalization of (3.42). On the other hand the orthogonality relation (3.43) is not valid in general with the simple weight function $e^{-x}$:

$$\int_0^\infty e^{-x} L_\alpha(x) L_\beta(x) dx \neq \delta_{\alpha\beta} \tag{3.47}$$

Hence the concept of orthogonality within the framework of fractional order polynomials is still an open question and needs further investigations [Abbott(2000)].

But it is indeed remarkable, that many well known formulas generating orthogonal polynomials may be extended using similar techniques as proposed for a generalization of the fractional derivative, e.g. the integral representation of Hermite polynomials $H_n$

$$H_n(x) = e^{x^2} \frac{2^{n+1}}{\sqrt{\pi}} \int_0^\infty e^{-t^2} t^n \cos\left(2xt - \frac{\pi}{2} n\right) dt \quad , n \in \mathbb{N} \tag{3.48}$$

may be extended to a definition of fractional order Hermite polynomials $H_\alpha$ via

$$H_\alpha(x) = e^{x^2} \frac{2^{\alpha+1}}{\sqrt{\pi}} \int_0^\infty e^{-t^2} t^\alpha \cos\left(2xt - \frac{\pi}{2} \alpha\right) dt \quad , \alpha \in \mathbb{C} \tag{3.49}$$

The recurrence relations for Legendre polynomials $P_n$:

$$(n+1) P_{n+1}(x) = (2n+1) x P_n(x) - n P_{n-1}(x) \quad , n \in \mathbb{N} \tag{3.50}$$

remains valid for $P_\alpha$ if we extend the definition for $n \in \mathbb{N}$

$$P_n(x) = {}_2F_1\left(-n, n+1; 1; \frac{1-x}{2}\right) \tag{3.51}$$

to $\alpha \in \mathbb{C}$

$$P_\alpha(x) = {}_2F_1\left(-\alpha, \alpha+1; 1; \frac{1-x}{2}\right) \tag{3.52}$$

and is then given by:

$$(\alpha+1) P_{\alpha+1}(x) = (2\alpha+1) x P_\alpha(x) - \alpha P_{\alpha-1}(x) \quad , \alpha \in \mathbb{C} \tag{3.53}$$

### 3.3.2 *Differential representation of the Riemann fractional derivative*

**Question:**
You have proposed a series expansion as a strategy to extend the scope of the fractional derivative types presented. Will the fractional Leibniz product rule serve as a promising tool to determine the exact form of such an extension?

**Answer:**
Yes, the fractional Leibniz product rule (3.39) allows for an appropriate generalization of a given fractional derivative definition for a special function class and leads to a differential representation of this general definition. Starting e.g. with the Riemann–Euler definition of a fractional derivative of $x^\beta$ and using the fractional extension of the Leibniz product rule indeed leads to a series expansion of the Riemann fractional derivative which is valid for analytic functions $f(x)$.

To derive the explicit form, we write the analytic function $f(x)$ as a general product:

$$f(x) = x^\beta g(x), \qquad \beta, x \geq 0 \tag{3.54}$$

We now apply the Leibniz product rule

$$\begin{aligned}
\partial_x^\alpha f(x) &= \partial_x^\alpha (x^\beta g(x)) \\
&= \sum_{j=0}^{\infty} \binom{\alpha}{j} (\partial_x^{\alpha-j} x^\beta)(\partial_x^j g(x)) \\
&= \sum_{j=0}^{\infty} \binom{\alpha}{j} \frac{\Gamma(1+\beta)}{\Gamma(1+\beta-\alpha+j)} x^{\beta-\alpha+j} (\partial_x^j g(x)) \\
&= \Gamma(1+\beta) x^{\beta-\alpha} \sum_{j=0}^{\infty} \binom{\alpha}{j} \frac{1}{\Gamma(1+\beta-\alpha+j)} x^j (\partial_x^j g(x)) \\
&= \Gamma(1+\beta) x^{\beta-\alpha} \sum_{j=0}^{\infty} (-1)^j \binom{j-\alpha-1}{j} \frac{1}{\Gamma(1+\beta-\alpha+j)} x^j (\partial_x^j g(x))
\end{aligned} \tag{3.55}$$

We now introduce the Euler-operator $J_e = x\partial_x$ and define the symbol :: as a normal order operator

$$: J_e^n := : (x\partial_x)^n := x^n \partial_x^n \tag{3.56}$$

It then follows

$$\begin{aligned}
\partial_x^\alpha f(x) &= \Gamma(1+\beta)x^{\beta-\alpha}\sum_{j=0}^{\infty}\binom{j-\alpha-1}{j}\frac{1}{\Gamma(1+\beta-\alpha+j)} :(-x\partial_x)^j: g(x) \\
&= \Gamma(1+\beta)x^{\beta-\alpha}\sum_{j=0}^{\infty}\frac{\Gamma(j-\alpha)}{\Gamma(-\alpha)\Gamma(1+j)\Gamma(1+\beta-\alpha+j)} :(-J_e)^j: g(x) \\
&= \frac{\Gamma(1+\beta)}{\Gamma(1+\beta-\alpha)}x^{\beta-\alpha}\sum_{j=0}^{\infty}\frac{\Gamma(j-\alpha)}{\Gamma(-\alpha)}\frac{\Gamma(1+\beta-\alpha)}{\Gamma(1+\beta-\alpha+j)}\frac{1}{j!} :(-J_e)^j: g(x) \\
&= \frac{\Gamma(1+\beta)}{\Gamma(1+\beta-\alpha)}x^{\beta-\alpha} :{}_1F_1(-\alpha;1+\beta-\alpha;-J_e): g(x) \qquad (3.57)
\end{aligned}$$

For $\beta = 0$ this reduces to

$$\partial_x^\alpha f(x) = \frac{1}{\Gamma(1-\alpha)}x^{-\alpha} :{}_1F_1(-\alpha;1-\alpha;-J_e): f(x) \quad x \geq 0 \qquad (3.58)$$

Interpreting the hypergeometric function ${}_1F_1$ as a series expansion, equation (3.58) represents the differential representation of the Riemann fractional derivative for an analytic function, as long as this series is convergent. It should be noted, that the knowledge of all derivatives $\partial_x^n f(x)$ is necessary to obtain the exact result. This amount of necessary information is equivalent to a full Taylor series expansion of the given function. This is a first hint for the nonlocality of the fractional derivative.

Furthermore equation (3.58) may be intuitively extended to negative x-values. We propose as a possible generalization:

$$\partial_x^\alpha f(x) = \frac{1}{\Gamma(1-\alpha)}\text{sign}(x)|x|^{-\alpha} :{}_1F_1(-\alpha;1-\alpha;-J_e): f(x) \quad x \in \mathbb{R} \qquad (3.59)$$

where $|x|$ denotes the absolute value of $x$.

As an example, the fractional derivative of the exponential function according to the Riemann definition is given as

$$\begin{aligned}
\partial_x^\alpha e^{kx} &= \frac{1}{\Gamma(1-\alpha)}\text{sign}(x)|x|^{-\alpha}{}_1F_1(-\alpha;1-\alpha;-kx)\, e^{kx} \qquad (3.60) \\
&= \frac{1}{\Gamma(1-\alpha)}\text{sign}(x)|x|^{-\alpha}{}_1F_1(1;1-\alpha;kx) \qquad (3.61) \\
&= \text{sign}(x)(\text{sign}(x)k)^\alpha e^{kx}\left(1-\frac{\Gamma(-\alpha,kx)}{\Gamma(-\alpha)}\right) \qquad (3.62)
\end{aligned}$$

where $\Gamma(a,z)$ is the incomplete gamma function. The result is valid and well behaved for all $k, x \in \mathbb{R}$.

As a remarkable fact, we have used methods, which were available already in the first half of the eighteenth century [Euler(1738)] to derive a differential form of the generalized Riemann fractional derivative, which did not play any role in the historic development of fractional calculus, while the integral form of the same fractional derivative was derived not until the mid-nineteenth century by Riemann [Riemann(1847)].

# Chapter 4

# Friction Forces

*To start with a practical application of the fractional calculus on problems in the area of classical physics we will investigate the influence of friction forces on the dynamical behaviour of classical particles. Within the framework of Newtonian mechanics friction phenomena are treated mostly schematically. Therefore a description using fractional derivatives may lead to results of major importance.*

*There are two reasons to start with this subject. This is a nice example, where fractional calculus may be applied on a very early entry level in physics. In addition, for a thorough treatment of fractional friction forces it is not necessary to use an exceptionally difficult mathematical framework, the hitherto presented mathematical tools are sufficient. It is also the first example, that an analytic approach using the methods developed in fractional calculus leads to results, which in general reach beyond the scope of a classical treatment. More examples will be presented in the following chapters.*

*First we will recapitulate the classical approach and will then discuss the solutions of Newton's equations of motion including fractional friction.*

*It will be demonstrated, that the viewpoint of fractional calculus leads to new insights and surprising interrelations of classical fields of research that remain unconnected until now.*

## 4.1 Classical description

We owe Sir Isaac Newton the first systematic application of differential calculus. Newton's mechanics abstracts from real complex physical processes introducing an idealized dimensionless point mass. Its dynamical behaviour is completely specified giving the position $x(t)$ and the velocity

$v(t) = \dot{x}(t) = \frac{d}{dt}x(t)$ as a function of time $t$.

The dynamic development of these quantities caused by an external force $F$ is determined by Newton's second law:

$$F = ma = m\ddot{x}(t) = m\frac{d^2}{dt^2}x(t) \tag{4.1}$$

This is an ordinary second order differential equation for $x(t)$ depending on the time variable $t$. In absence of external forces $F = 0$ the general solution of this equation follows from twofold integration of $\ddot{x} = 0$:

$$x(t) = c_1 + c_2 t \tag{4.2}$$

with two at first arbitrary constants of integration $c_1$ and $c_2$, which may be determined defining appropriately chosen initial conditions.

For a given initial position $x_0$ and initial velocity $v_0$

$$\begin{aligned} x(t = 0) &= x_0 \\ v(t = 0) &= v_0 \end{aligned} \tag{4.3}$$

it follows

$$c_1 = x_0 \tag{4.4}$$

$$c_2 = v_0 \tag{4.5}$$

and therefore

$$x(t) = x_0 + v_0 t \tag{4.6}$$

$$\dot{x}(t) = v_0 \tag{4.7}$$

Hence we conclude, that a point mass keeps its initial velocity forever in absence of external forces. This statement is known as Newton's first law.

The observation of everyday motion contradicts this result. Sooner or later every kind of motion comes to rest without outer intervention. A wheel stops spinning. The ball will not roll any more, but will lie still in the grass. If we stop rowing, a boat will not glide forward any more.

All these observed phenomena have different physical causes. Within the framework of Newton's theory they are summarized as friction forces $F_R$. To be a little bit more specific we consider as friction forces all kinds of forces which point in the opposite direction of the velocity of a particle.

Therefore an ansatz for friction forces is a simple power law:

$$F_R = -\mu\,\text{sign}(v)|v|^\alpha \tag{4.8}$$

with an arbitrarily chosen real exponent $\alpha$.

The following special cases are:

- $\alpha \approx 0$ is observed for static and kinetic friction for solids.
- $\alpha = 1$ Stokes friction in liquids with high viscosity.
- $\alpha = 2$ is a general trend for high velocities .

In reality both gases and liquids show a behaviour which only approximately corresponds to these special cases. For a more realistic treatment a superposition of these different cases leads to better results.

In the following we want to discuss the solutions for an equation of motion including the friction force of type (4.8). We have to solve the following differential equation:

$$m\ddot{x} = -\mu \, \text{sign}(v)|\dot{x}|^{\alpha} \tag{4.9}$$

Within the mks unit-system the mass $m$ is measured in $[kg]$ and the friction coefficient $\mu$ is given as $[kg/s^{2-\alpha}]$. Assuming $v(t) > 0$ equation (4.9) reduces to

$$m\ddot{x} = -\mu\dot{x}^{\alpha}, \qquad \dot{x} > 0 \tag{4.10}$$

Despite the fact that this is a nonlinear differential equation, it may be solved directly with an appropriately chosen ansatz. We start with:

$$x(t) = c_1 + c_2(1 + bt)^a \tag{4.11}$$

Using the initial conditions (4.3) we obtain the general solution for (4.10):

$$x(t) = x_0 + \frac{mv_0^{2-\alpha}}{\mu(2-\alpha)} - \frac{m\left((\alpha-1)(-\frac{v_0^{1-\alpha}}{1-\alpha} + \frac{\mu}{m}t)\right)^{\frac{2-\alpha}{1-\alpha}}}{\mu(2-\alpha)} \tag{4.12}$$

This equation determines the motion of a point mass influenced by a velocity dependent friction force with arbitrary $\alpha$. The well known special cases $\alpha = 0$, $\alpha = 1$ and $\alpha = 2$ are included as limiting cases:

$$\lim_{\alpha\to 0} x(t) = x_0 + v_0 t - \frac{1}{2}\frac{\mu}{m}t^2 \tag{4.13}$$

$$\lim_{\alpha\to 1} x(t) = x_0 + \frac{m}{\mu}v_0(1 - e^{-\frac{\mu}{m}t}) \tag{4.14}$$

$$\lim_{\alpha\to 2} x(t) = x_0 + \frac{m}{\mu}\log\left(1 + \frac{\mu}{m}v_0 t\right) \tag{4.15}$$

A series expansion of (4.12) up to second order in $t$ results in:

$$x(t) \approx x_0 + v_0 t - \frac{1}{2}\frac{\mu}{m}v_0^{\alpha}t^2 \tag{4.16}$$

Therefore the point mass is object to a negative acceleration or deceleration proportional to $v_0^{\alpha}$.

## 4.2 Fractional friction

We will now discuss friction phenómena within the framework of fractional calculus. According to the classical results presented in the previous section we propose the following fractional friction force:

$$F_R = -\mu \frac{d^\alpha}{dt^\alpha} x(t) \qquad \dot{x}(t) > 0 \tag{4.17}$$

Here we have introduced a fractional derivative coefficient $\alpha$ which is an arbitrary real number and the fractional friction coefficient $\mu$ in units $[kg/s^{2-\alpha}]$ which parametrizes the strength of the fractional friction force.

For the special case $\alpha = 1$ classical and fractional friction force coincide. But for all other cases we have

$$\dot{x}^\alpha = \left(\frac{d}{dt}x(t)\right)^\alpha \neq \frac{d^\alpha}{dt^\alpha}x(t) \tag{4.18}$$

and hence we expect a dynamic behaviour which differs from the classical result. We define the following fractional differential equation

$$m\ddot{x} = -\mu \frac{d^\alpha}{dt^\alpha} x(t) \qquad \dot{x}(t) > 0 \tag{4.19}$$

with initial conditions (4.3), which determines the dynamical behaviour of a classical point mass object to a fractional friction force.

A solution of this equation may be realized with the ansatz

$$x(t) = e^{\omega t} \tag{4.20}$$

and applying the Liouville definition of the fractional derivative

$$\frac{d^\alpha}{dt^\alpha} e^{\omega t} = \omega^\alpha e^{\omega t} \tag{4.21}$$

Inserting this ansatz into the differential equation the general solution of the equation may be obtained, if the complete set of solutions for the polynomial

$$\omega^2 = -\frac{\mu}{m}\omega^\alpha \tag{4.22}$$

is determined.

The trivial solution $\omega = 0$ for $\alpha > 0$ describes a constant, time independent function $x(t) = \text{const}$ and is a consequence of the fact, that the differential equation is built from derivatives of $x(t)$ only.

Additional solutions of the polynomial equation are formally given by

$$\omega = (-1)^{\frac{1}{2-\alpha}} \left|\frac{\mu}{m}\right|^{\frac{1}{2-\alpha}} \tag{4.23}$$

The actual number of different solutions depends on the numerical value of $\alpha$.

If $\alpha$ is in the range of $0 < \alpha < 1$ two different conjugate complex solutions exist. Using the abbreviation

$$\kappa = \left|\frac{\mu}{m}\right|^{\frac{1}{2-\alpha}} \tag{4.24}$$

these solutions are given as:

$$\omega_1 = \kappa \left(\cos\left(\frac{\pi}{2-\alpha}\right) + i\sin\left(\frac{\pi}{2-\alpha}\right)\right) \tag{4.25}$$

$$\omega_2 = \kappa \left(\cos\left(\frac{\pi}{2-\alpha}\right) - i\sin\left(\frac{\pi}{2-\alpha}\right)\right) \tag{4.26}$$

Hence we obtain the general solution of the fractional differential equation (4.19):

$$x(t) = c_1 + c_2 e^{\omega_1 t} + c_3 e^{\omega_2 t} \quad 0 < \alpha < 1 \tag{4.27}$$

A first remarkable result of our investigation is the fact, that there are three different constants $c_i$, where two of these may be determined independently using the initial conditions (4.3). Consequently we are free to specify an additional reasonable initial condition. This clearly indicates that the fractional differential equation (4.19) describes a broader range of phenomena than the classical equation of motion.

One possible choice for an additional initial condition results from the series expansion (4.16) of the equation of motion including classical friction. For the second derivative of the fractional differential equation we demand:

$$\ddot{x}(t=0) = -\frac{\mu}{m} v_0^\alpha \tag{4.28}$$

This leads to the following set of determining equations for the three different coefficients $c_i$:

$$\begin{pmatrix} 1 & 1 & 1 \\ 0 & \omega_1 & \omega_2 \\ 0 & \omega_1^2 & \omega_2^2 \end{pmatrix} \begin{pmatrix} c_1 \\ c_2 \\ c_3 \end{pmatrix} = \begin{pmatrix} x_0 \\ v_0 \\ -\frac{\mu}{m} v_0^\alpha \end{pmatrix} \tag{4.29}$$

As a consequence the solutions of the fractional differential equation agree up to second order in $t$ with solutions obtained for the Newtonian equation of motion including the classical friction term (4.8).

$$x(t) \approx x_0 + v_0 t - \frac{1}{2}\frac{\mu}{m} v_0^\alpha t^2 + o(t^3) \tag{4.30}$$

From Fig. 4.1 we may deduce, that indeed the difference of classical and fractional solution becomes manifest only for $t \gg 1$.

Remarkably enough there is an alternative choice for the additional initial condition. Let us assume that the initial velocity of a mass point decelerates under the influence of a fractional friction force. In other words during all stages of motion the initial velocity is maximal and therefore at least a local extremum of the velocity graph.

Besides the initial conditions (4.3) we therefore define as an additional condition:

$$\dot{v}(t=0) = \ddot{x}(t=0) = 0 \tag{4.31}$$

This leads to the following set of determining equations for the coefficients $c_i$:

$$\begin{pmatrix} 1 & 1 & 1 \\ 0 & \omega_1 & \omega_2 \\ 0 & \omega_1^2 & \omega_2^2 \end{pmatrix} \begin{pmatrix} c_1 \\ c_2 \\ c_3 \end{pmatrix} = \begin{pmatrix} x_0 \\ v_0 \\ 0 \end{pmatrix} \tag{4.32}$$

The complete solution of the fractional differential equation including a fractional friction force is then given by:

$$\begin{aligned} x(t) &= x_0 \\ &\quad + \frac{v_0}{\kappa \sin(\frac{\pi}{2-\alpha})} \left( e^{\kappa \cos(\frac{\pi}{2-\alpha})\, t} \sin\left(\frac{2\pi}{2-\alpha} - \kappa \sin\left(\frac{\pi}{2-\alpha}\right) t\right) \right. \\ &\quad \left. - \sin\left(\frac{2\pi}{2-\alpha}\right)\right) \end{aligned} \tag{4.33}$$

This solution is valid from $t = 0$ up to a final time $t_0$ where the motion of the mass point comes to rest and then remains at the final position. $t_0$ is determined by:

$$t_0 = \frac{\pi}{(2-\alpha)\kappa \sin(\frac{\pi}{2-\alpha})} \tag{4.34}$$

Special cases of the general solution (4.33) are solutions for $\alpha = 0$ and $\alpha = 1$ respectively:

$$x(t, \alpha = 0) = x_0 + \frac{v_0}{\kappa} \sin(\kappa t) \tag{4.35}$$

$$x(t, \alpha = 1) = x_0 + \frac{v_0}{\kappa} \left(2 - 2e^{-\kappa t} - \kappa\, t\, e^{-\kappa t}\right) \tag{4.36}$$

Obviously this is a completely new function type. The initial velocity $v_0$ becomes less important and merely acts as a scaling parameter. The time evolution of the solutions may be understood much better in terms of free ($\alpha = 0$) and damped ($0 < \alpha \leq 1$) oscillations, if we consider the first quarter period for a description of a fractional friction process.

In order to investigate this aspect of fractional friction, for reasons of completeness we first present the solutions of the classical harmonic oscillator.

The differential equation for the damped harmonic oscillator is given by:

$$m\ddot{x}(t) = -kx(t) - \gamma\dot{x}(t) \tag{4.37}$$

For a mass $m$ with the damping coefficient $\gamma$ and the spring constant $k$. The ansatz

$$x(t) = e^{\omega t} \tag{4.38}$$

leads to a quadratic equation for the frequencies $\omega$:

$$\omega^2 = -\frac{\gamma}{m}\omega - \frac{k}{m} \tag{4.39}$$

which leads to two solutions:

$$\omega_{1,2} = -\frac{\gamma}{2m} \pm \sqrt{\frac{\gamma^2}{4m^2} - \frac{k}{m}} \tag{4.40}$$

Depending on the value of $\gamma$ and of the discriminant $D = \frac{\gamma^2}{4m^2} - \frac{k}{m}$ we distinguish the cases:

- $\gamma = 0$ leads to a free oscillation.
- $\gamma > 0$ and $D < 0$ leads to a damped oscillation.
- $\gamma > 0$ and $D = 0$ There is no oscillatory contribution any more. The system returns to equilibrium as quickly as possible. This case is called the critically damped case.
- $\gamma > 0$ and $D > 0$ In this case too there is no oscillatory component any more. The system exponentially decays. This case is called the over damped case. The equilibrium position is reached for $t \to \infty$.

Now we can compare the frequencies $\omega_i$ from the fractional differential equation (4.25) to the frequencies for the damped harmonic oscillator (4.40).

We obtain the following relations for the parameter set $\{\gamma, k\}$ which determines the behaviour of the classical damped harmonic oscillator and the parameter set $\{\alpha, \mu\}$, which describes the behaviour of the solutions of the fractional friction differential equation:

$$\frac{k}{m} = \left(\frac{\mu}{m}\right)^{\frac{2}{2-\alpha}} \tag{4.41}$$

$$\frac{\gamma}{m} = -2\left(\frac{\mu}{m}\right)^{\frac{1}{2-\alpha}} \cos\left(\frac{\pi}{2-\alpha}\right) \tag{4.42}$$

From the above follows a unique relation between $\gamma/\sqrt{km}$ and the fractional derivative coefficient $\alpha$:

$$\frac{\gamma}{\sqrt{km}} = -2\cos\left(\frac{\pi}{2-\alpha}\right) \tag{4.43}$$

$$\approx \frac{\pi}{2}\alpha \qquad \text{for} \quad \alpha \approx 0 \tag{4.44}$$

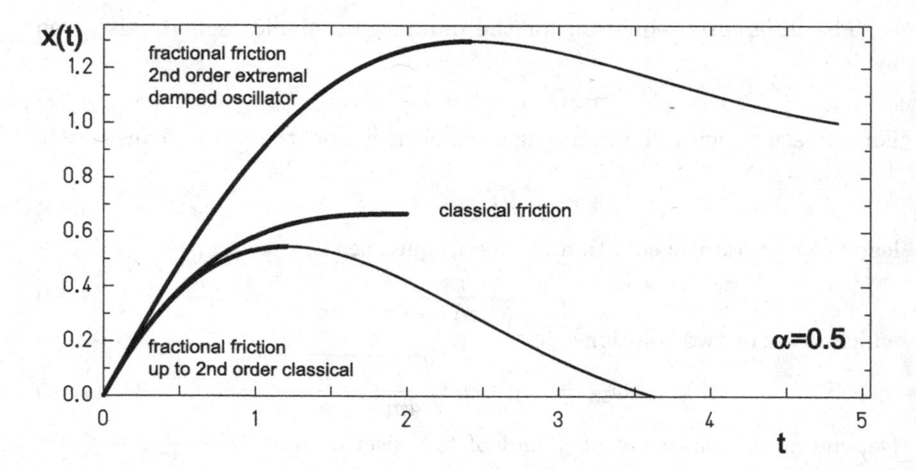


Fig. 4.1. For $\alpha = 0.5$ different solutions of equations of motion are plotted. In the middle the classical solution of the Newtonian differential equation (4.12) is presented, below the corresponding fractional solution with the additional initial condition (4.28) is shown. Up to second order in $t$ both solutions coincide. A difference may be observed only at the end of the motion period. The upper line indicates the fractional solution using the additional initial condition (4.31). Thick lines describe the motion valid until the body comes to rest. In addition, for both fractional solutions the plot range is extended to t-values $t > t_0$ (thin lines), which is beyond the physically valid region.

We have derived a first physical interpretation for the at first abstractly introduced fractional derivative coefficient $\alpha$. Obviously the ratio $\gamma/\sqrt{km}$ and the fractional derivative coefficient $\alpha$ are directly correlated.

We may conclude, that fractional friction may be interpreted in the limit $\alpha \to 0$ in analogy to the harmonic oscillator more and more as a damping term.

On the other hand, if we examine the region $\alpha \approx 2$ the fractional friction force may be understood as an additional contribution to the acceleration term. For the ideal case $\alpha = 2$ we obtain the differential equation

$$(m + \mu)\ddot{x} = 0 \tag{4.45}$$

which obviously describes an increase of mass. The point mass moves freely with an increased mass $m' = m + \mu$.

The variation of the fractional friction coefficient $\alpha$ within the interval $2 \geq \alpha \geq 0$ therefore describes acceleration- velocity- and position dependent classical force types from a generalized point of view as fractional friction.

We want to emphasize two aspects of our investigations performed so far.

Using the fractional equivalent to the Newtonian equation of motion with a classical friction term, we may describe a much broader area of application than it is possible with the classical approach. For an appropriately chosen set of fractional initial conditions we are able to describe damped oscillations. From the point of view of fractional calculus the harmonic oscillator is just one more example for a fractional friction phenomenon. In other words, while in a classical picture friction and damping are described using different approaches, the fractional picture allows for an unified description.

This is a general aspect of a fractional approach. Different phenomena, which seem to be unrelated in a classical description, now appear as different realizations of a general aspect in a fractional theory. This leads to intriguing, sometimes surprising new insights. We will demonstrate, that this is a common aspect of all applications of fractional calculus, as we will verify in the following chapters.

One more important observation is a rather technical remark. The fractional differential equation (4.19) merges two different classical types of a differential equation. On the one hand the solutions of a non linear differential equation, which is in general a quite complicated type of equation and on the other hand solutions of a simple second order differential equation with constant coefficients result from the same fractional equation.

This aspect is of general nature and nourishes the hope, that a wide range of complex problems may be solved analytically with minimal effort using fractional calculus, while a solution within the framework of a classical theory may be difficult. In the following chapters we will present some examples, to illustrate this point.

# Chapter 5

# Fractional Calculus

*Up to now we have introduced a fractional derivative definition for special simple function classes. In the following section we will present common generalizations for arbitrary analytic functions. As a starting point we will use Cauchy's formula for repeated integration which will be extended to fractional integrals. Having derived a fractional integral $I^\alpha$ properly we will obtain a satisfying definition of a fractional derivative, if we assume*

$$\frac{d^\alpha}{dx^\alpha} = I^{-\alpha} \tag{5.1}$$

*which is equivalent to demanding that a fractional integral is the inverse of a fractional derivative.*

*As a consequence, the complexity level and difficulties of calculating the fractional derivative for a given function will turn out to be similar to integral calculus. This is in contrast to a standard classical derivative, which, in general, is much easier to calculate than a corresponding integral.*

*First we will discuss possible generalizations for the first derivative, namely, the Liouville, Liouville–Caputo, Riemann- and Caputo fractional derivative definition on the basis of the different possible sequences of applying operators for the fractional integral and the derivative. This approach has the advantage, that it may be easily extended to more complex operator types, e.g. space-dependent derivatives.*

*We will also show, that higher order derivatives may be directly generalized on the basis of the Riesz-definition of the fractional derivative.*

*We will present Podlubny's [Podlubny(2002)] geometric interpretation of a fractional integral and will demonstrate, that we may establish a smooth transition from standard derivative ($\alpha = n$) via low level fractionality ($\alpha = n + \epsilon$) up to arbitrary fractional derivative following an idea proposed by Tarasov [Tarasov(2006)].*

*Finally we will present a derivation of the semigroup property*

$$I^{\alpha} I^{\beta} f(x) = I^{\alpha+\beta} f(x) \qquad \alpha, \beta \geq 0 \tag{5.2}$$

*which is the defining property for a non-local derivative operator of fractional type.*

## 5.1 The Fourier transform

A direct approach for a general fractional derivative had been proposed by Fourier in the early 19th century [Fourier(1822)]. Therefore we will present his idea first:

In the interval $[-L/2 \leq x \leq L/2]$ a function may be expanded in a Fourier series:

$$f(x) = \frac{1}{\sqrt{L}} \sum_{k=-\infty}^{k=+\infty} g_k e^{ik\frac{2\pi}{L}x} \tag{5.3}$$

where the complex valued coefficients $g_k$ are given by

$$g_k = \frac{1}{\sqrt{L}} \int_{-L/2}^{+L/2} f(x) e^{ik\frac{2\pi}{L}x} dx \tag{5.4}$$

Extending the interval length to infinity, the discrete spectrum $g_k$ changes to a continuous one $g(k)$.

The Fourier transform $g(k)$ is then given by

$$g(k) = \frac{1}{\sqrt{2\pi}} \int_{-\infty}^{+\infty} f(x) e^{ikx} dx \tag{5.5}$$

and the inverse is given as

$$f(x) = \frac{1}{\sqrt{2\pi}} \int_{-\infty}^{+\infty} g(k) e^{-ikx} dk \tag{5.6}$$

The derivative of a function results as

$$\frac{d^n}{dx^n} f(x) = \frac{1}{\sqrt{2\pi}} \int_{-\infty}^{+\infty} g(k)(-ik)^n e^{-ikx} dk \tag{5.7}$$

This result may be directly extended to a fractional derivative definition:

$$\frac{d^{\alpha}}{dx^{\alpha}} f(x) = \frac{1}{\sqrt{2\pi}} \int_{-\infty}^{+\infty} g(k)(-ik)^{\alpha} e^{-ikx} dk \tag{5.8}$$

Equations (5.5) and (5.8) define the fractional derivative according to Fourier. This definition is quite simple but very elegant. The only condition is the existence of the Fourier transform $g(k)$ and its fractional inverse

Table 5.1. Some special functions and their fractional derivative according to Fourier

| $f(x)$ | $\frac{d^\alpha}{dx^\alpha} f(x)$ |
|---|---|
| $e^{ikx}$ | $(ik)^\alpha e^{ikx}$ |
| $\sin(kx)$ | $k^\alpha \sin(kx + \frac{\pi}{2}\alpha)$ |
| $\cos(kx)$ | $k^\alpha \cos(kx + \frac{\pi}{2}\alpha)$ |
| $e^{-kx^2}$ | $\frac{2^\alpha}{\sqrt{\pi}} k^{\alpha/2} \big(\cos(\frac{\pi}{2}\alpha)\Gamma(\frac{1+\alpha}{2})_1F_1(\frac{1+\alpha}{2}; \frac{1}{2}; -kx^2)$ $-\sqrt{k}x\alpha \sin(\frac{\pi}{2}\alpha)\Gamma(\frac{\alpha}{2})_1F_1(\frac{2+\alpha}{2}; \frac{3}{2}; -kx^2)\big)$ |
| $\mathrm{erf}(kx)$ | $\frac{2^\alpha}{\sqrt{\pi}} k^\alpha \big(2kx \cos(\frac{\pi}{2}\alpha)\Gamma(\frac{1+\alpha}{2})_1F_1(\frac{1+\alpha}{2}; \frac{3}{2}; -k^2x^2)$ $+ \sin(\frac{\pi}{2}\alpha)\Gamma(\frac{\alpha}{2})_1F_1(\frac{\alpha}{2}; \frac{1}{2}; -k^2x^2)\big)$ |
| $\log(x)$ | $\frac{\Gamma(\alpha)}{2\|x\|^\alpha} \left(e^{i\pi\alpha} + 1 + \mathrm{sign}(x)(e^{i\pi\alpha} - 1)\right)$ |
| $\|x\|^{-k}$ | $\frac{\Gamma(k+\alpha)}{\Gamma(k)} \|x\|^{-k-\alpha}, \quad x < 0$ |

for a given function. In addition, the fractional derivative parameter $\alpha$ may be chosen arbitrarily.

In particular the real part of $\alpha$ may also be a negative number. In this case we call (5.8) a fractional integral definition.

In Table 5.1 we have listed some special functions and the corresponding fractional derivative according to Fourier.

## 5.2 The fractional integral

If we use the abbreviation ${}_aI$ for the integral operator

$$ {}_aIf = \int_a^x f(\xi) d\xi \tag{5.9} $$

integration of a function may be considered as the inverse operation of differentiation:

$$ \left(\frac{d}{dx}\right)({}_aI)f = f \tag{5.10} $$

Hence it is reasonable to investigate the method of fractional integration first and based on this investigation in a next step to make statements on the fractional derivative of analytic functions in general.

Let us start with the observation, that the multiple integral

$$ {}_aI^n f = \int_a^{x_n} \int_a^{x_{n-1}} \ldots \int_a^{x_1} f(x_0) dx_0 \ldots dx_{n-1} \tag{5.11} $$

using Cauchy's formula of repeated integration may be reduced to a single integral

$$ {}_aI^n f(x) = \frac{1}{(n-1)!} \int_a^x (x - \xi)^{n-1} f(\xi) d\xi \tag{5.12} $$

This formula may be easily extended to the fractional case:

$$ {}_aI_+^\alpha f(x) = \frac{1}{\Gamma(\alpha)} \int_a^x (x-\xi)^{\alpha-1} f(\xi) d\xi \tag{5.13}$$

$$ {}_bI_-^\alpha f(x) = \frac{1}{\Gamma(\alpha)} \int_x^b (\xi - x)^{\alpha-1} f(\xi) d\xi \tag{5.14}$$

The first of these two equations is valid for $x > a$, the second one is valid for $x < b$. The constants $a, b$ determine the lower and upper boundary of the integral domain and may at first be arbitrarily chosen. To distinguish the two cases, we call them the left- and right-handed case respectively. This may be understood from a geometrical point of view. The left-handed integral collects weighted function values for $\xi < x$, which means left from x. The right-handed integral collects weighted function values for $\xi > x$, which means right from x. If x is a time-like coordinate, then the left-handed integral is causal, the right handed integral is anti-causal.

The actual value of the integral obviously depends on the specific choice of the two constants $(a, b)$.

This will indeed be the remarkable difference for the two mostly used different definitions of a fractional integral:

### 5.2.1 *The Liouville fractional integral*

We define the fractional integral according to Liouville setting $a = -\infty$, $b = +\infty$:

$$ {}_{\mathrm{L}}I_+^\alpha f(x) = \frac{1}{\Gamma(\alpha)} \int_{-\infty}^x (x-\xi)^{\alpha-1} f(\xi) d\xi \tag{5.15}$$

$$ {}_{\mathrm{L}}I_-^\alpha f(x) = \frac{1}{\Gamma(\alpha)} \int_x^{+\infty} (\xi - x)^{\alpha-1} f(\xi) d\xi \tag{5.16}$$

### 5.2.2 *The Riemann fractional integral*

The fractional integral according to Riemann is given by setting $a = 0$, $b = 0$:

$$ {}_{\mathrm{R}}I_+^\alpha f(x) = \frac{1}{\Gamma(\alpha)} \int_0^x (x-\xi)^{\alpha-1} f(\xi) d\xi \tag{5.17}$$

$$ {}_{\mathrm{R}}I_-^\alpha f(x) = \frac{1}{\Gamma(\alpha)} \int_x^0 (\xi - x)^{\alpha-1} f(\xi) d\xi \tag{5.18}$$

What is the difference between the definition according to Liouville (5.15), (5.16) and that according to Riemann (5.17), (5.18)? Let us apply both integral definitions to the special function $f(x) = e^{kx}$. We obtain:

$$
\begin{aligned}
{}_{\mathrm{L}}I_{+}^{\alpha} e^{kx} &= k^{-\alpha} e^{kx} && k, x > 0 && (5.19)\\
{}_{\mathrm{L}}I_{-}^{\alpha} e^{kx} &= (-k)^{-\alpha} e^{kx} && k < 0 && (5.20)\\
{}_{\mathrm{R}}I_{+}^{\alpha} e^{kx} &= k^{-\alpha} e^{kx} \left(1 - \frac{\Gamma(\alpha, kx)}{\Gamma(\alpha)}\right) && x > 0 && (5.21)\\
{}_{\mathrm{R}}I_{-}^{\alpha} e^{kx} &= (-k)^{-\alpha} e^{kx} \left(1 - \frac{\Gamma(\alpha, kx)}{\Gamma(\alpha)}\right) && x < 0 && (5.22)
\end{aligned}
$$

where $\Gamma(\alpha, x)$ is the incomplete gamma function.

Using the Liouville fractional integral definition the function $e^{kx}$ vanishes for $k > 0$ at the lower boundary of the integral domain. Consequently there is no additional contribution. On the other hand with the Riemann definition we obtain an additional contribution, since at $a = 0$ the exponential function does not vanish.

## 5.3 Correlation of fractional integration and differentiation

In the last section we have presented the Riemann and Liouville version of a fractional integral definition. But actually we are interested in a useful definition of a fractional derivative. Using the abbreviation

$$
\frac{d^{\alpha}}{dx^{\alpha}} = D^{\alpha} \tag{5.23}
$$

we will demonstrate, that the concepts of fractional integration and fractional differentiation are closely related. Let us split up the fractional derivative operator

$$
\begin{aligned}
D^{\alpha} &= D^{m} D^{\alpha - m} \qquad m \in \mathbb{N} && (5.24)\\
&= \frac{d^{m}}{dx^{m}}\, {}_{a}I^{m-\alpha} && (5.25)
\end{aligned}
$$

This means that a fractional derivative may be interpreted as a fractional integral followed by an ordinary standard derivative. Once a definition of the fractional integral is given the fractional derivative is determined too.

The inverted sequence of operators

$$D^{\alpha} = D^{\alpha-m} D^{m} \qquad m \in \mathbb{N} \tag{5.26}$$

$$= {}_{a}I^{m-\alpha} \frac{d^{m}}{dx^{m}} \tag{5.27}$$

leads to an alternative decomposition of the fractional derivative into an ordinary standard derivative followed by a fractional integral.

Both decompositions (5.24) and (5.26) of course lead to different results.

With these decompositions we are able to understand the mechanism, how nonlocality enters the fractional calculus. The standard derivative of course is a local operator, but the fractional integral certainly is not. The fractional derivative must be understood as the inverse of fractional integration, which is a nonlocal operation. As a consequence, fractional differentiation and fractional integration yield the same level of difficulty.

In the last section we have given two different definitions of the fractional integral from Liouville and Riemann. For each of these definitions according to the two different decompositions given above follow two different realizations of a fractional derivative definition, which we will present in the following.

### 5.3.1 *The Liouville fractional derivative*

For the simple case $0 < \alpha < 1$ we obtain for the Liouville definition of a fractional derivative using (5.15),(5.16) and operator sequence (5.24)

$$_{\mathrm{L}}D_{+}^{\alpha} f(x) = \frac{d}{dx} {}_{\mathrm{L}}I_{+}^{1-\alpha} f(x) \tag{5.28}$$

$$= \frac{d}{dx} \frac{1}{\Gamma(1-\alpha)} \int_{-\infty}^{x} (x-\xi)^{-\alpha} f(\xi) d\xi \tag{5.29}$$

$$_{\mathrm{L}}D_{-}^{\alpha} f(x) = \frac{d}{dx} {}_{\mathrm{L}}I_{-}^{1-\alpha} f(x) \tag{5.30}$$

$$= \frac{d}{dx} \frac{1}{\Gamma(1-\alpha)} \int_{x}^{+\infty} (\xi-x)^{-\alpha} f(\xi) d\xi \tag{5.31}$$

In Table 5.2 we have listed some special functions and their Liouville fractional derivative.

Table 5.2. The fractional derivative ${}_{\mathrm{L}}D^{\alpha}_{+}$ according to Liouville for some special functions

| $f(x)$ | $\frac{d^\alpha}{dx^\alpha} f(x)$ |
|---|---|
| $e^{kx}$ | $k^\alpha e^{kx} \qquad k \geq 0$ |
| $\sin(kx)$ | $k^\alpha \sin(kx + \frac{\pi}{2}\alpha)$ |
| $\cos(kx)$ | $k^\alpha \cos(kx + \frac{\pi}{2}\alpha)$ |
| $\text{erf}(kx)$ | divergent |
| $e^{-kx^2}$ | $\frac{k^{\frac{\alpha}{2}}}{\Gamma(1-\alpha)} \Big(\Gamma(1-\frac{\alpha}{2}) {}_1F_1(\frac{1}{2}+\frac{\alpha}{2};\frac{1}{2};-kx^2)$ $-\sqrt{k}\alpha x \Gamma(\frac{1}{2}-\frac{\alpha}{2}) {}_1F_1(1+\frac{\alpha}{2};\frac{3}{2};-kx^2)\Big)$ $-2\sqrt{k}\alpha x\Gamma(\frac{3}{2}-\frac{\alpha}{2}) {}_1F_1(\frac{1}{2}+\frac{\alpha}{2};\frac{3}{2};-kx^2)$ $-\frac{2}{3}k(1-\alpha^2)x^2\Gamma(\frac{1}{2}-\frac{\alpha}{2}) {}_1F_1(\frac{3}{2}+\frac{\alpha}{2};\frac{5}{2};-kx^2))$ |
| ${}_pF_q(\{a_i\};\{b_j\};kx)$ | $k^\alpha \prod_{i=1}^{p} \frac{\Gamma(a_i+\alpha)}{\Gamma(a_i)} \prod_{j=1}^{q} \frac{\Gamma(b_j)}{\Gamma(b_j+\alpha)}$ ${}_pF_q(\{a_i+\alpha\};\{b_j+\alpha\};kx)$ |
| $\lvert x\rvert^{-k}$ | $\frac{\Gamma(k+\alpha)}{\Gamma(k)} \lvert x\rvert^{-k-\alpha}, \qquad x < 0$ |

### 5.3.2 *The Riemann fractional derivative*

For the Riemann fractional derivative we obtain using (5.17), (5.18) and operator sequence (5.24):

$$
{}_{\mathrm{R}}D^{\alpha}_{+} f(x) = \frac{d}{dx} {}_{\mathrm{R}}I^{1-\alpha}_{+} f(x) \tag{5.32}
$$

$$
= \frac{d}{dx} \frac{1}{\Gamma(1-\alpha)} \int_0^x (x-\xi)^{-\alpha} f(\xi) d\xi \tag{5.33}
$$

$$
{}_{\mathrm{R}}D^{\alpha}_{-} f(x) = \frac{d}{dx} {}_{\mathrm{R}}I^{1-\alpha}_{-} f(x) \tag{5.34}
$$

$$
= \frac{d}{dx} \frac{1}{\Gamma(1-\alpha)} \int_x^0 (\xi-x)^{-\alpha} f(\xi) d\xi \tag{5.35}
$$

In Table 5.3 we have listed some special functions and their Riemann fractional derivative.

Table 5.3. Some special functions and their fractional derivative ${}_{\mathrm{R}}D^{\alpha}$ according to the Riemann definition

| $f(x)$ | $\frac{d^\alpha}{dx^\alpha} f(x)$ |
|---|---|
| $e^{kx}$ | $\operatorname{sign}(x)(\operatorname{sign}(x)k)^\alpha e^{kx}\left(1-\frac{\Gamma(-\alpha,kx)}{\Gamma(-\alpha)}\right)$ |
| $\sin(kx)$ | $\frac{(2-\alpha)\, k\operatorname{sign}(x)\lvert x\rvert^{-\alpha}x}{(2-3\alpha+\alpha^2)\Gamma(1-\alpha)}\, {}_1F_2(1;\frac{3}{2}-\frac{\alpha}{2},2-\frac{\alpha}{2};-\frac{1}{4}k^2x^2)$ $-\frac{k^3\operatorname{sign}(x)\lvert x\rvert^{-\alpha}x^3}{(\frac{3}{2}-\frac{\alpha}{2})(2-\frac{\alpha}{2})(2-3\alpha+\alpha^2)\Gamma(1-\alpha)}\, {}_1F_2(2;\frac{5}{2}-\frac{\alpha}{2},3-\frac{\alpha}{2};-\frac{1}{4}k^2x^2)$ |
| $\cos(kx)$ | $\operatorname{sign}(x)\frac{\lvert x\rvert^{-\alpha}}{\Gamma(4-\alpha)}$ $((\alpha-1)(\alpha-2)(\alpha-3)\,{}_1F_2(1;1-\frac{\alpha}{2},\frac{3}{2}-\frac{\alpha}{2};-\frac{1}{4}k^2x^2)$ $+2k^2x^2\,{}_1F_2(2;2-\frac{\alpha}{2},\frac{5}{2}-\frac{\alpha}{2};-\frac{1}{4}k^2x^2))$ |
| $\operatorname{erf}(kx)$ | $-2^{-1+\alpha}k\operatorname{sign}(x)\lvert x\rvert^{-\alpha}$ $\Big((a-2)\,{}_2\tilde{F}_2(\frac{1}{2},1;\frac{3}{2}-\frac{\alpha}{2},2-\frac{\alpha}{2};-k^2x^2)$ $+k^2x^2\,{}_2\tilde{F}_2(\frac{3}{2},2;\frac{5}{2}-\frac{\alpha}{2},3-\frac{\alpha}{2};-k^2x^2)\Big)$ |
| ${}_pF_q(\{a_i\};\{b_j\};kx)$ | $\operatorname{sign}(x)\lvert x\rvert^{-\alpha}\frac{1}{\Gamma(1-\alpha)}\,{}_{p+1}F_{q+1}(\{1,1+a_i\};\{b_j,2-\alpha\};kx)$ $+k\operatorname{sign}(x)\lvert x\rvert^{-\alpha}x\frac{1}{(1-\alpha)(2-\alpha)\Gamma(1-\alpha)}\prod_{i=1}^{p}a_i\prod_{j=1}^{q}\frac{1}{b_j}$ $\times\,{}_{p+1}F_{q+1}(\{2,1+a_i\};\{1+b_j,3-\alpha\};kx)$ |
| $\log(x)$ | $\frac{x^{-\alpha}}{\Gamma(2-\alpha)}\left(1-(1-\alpha)(H_{1-\alpha}+\log(x))\right), \quad x>0$ |
| $x^k$ | $\frac{\Gamma(1+k)}{\Gamma(1+k-\alpha)}\operatorname{sign}(x)\lvert x\rvert^{-\alpha}x^k$ |

### 5.3.3 *The Liouville fractional derivative with inverted operator sequence: the Liouville–Caputo fractional derivative*

In case we invert the operator sequence according to (5.26) we obtain on the basis of the Liouville fractional integral definition (5.15), (5.16):

$$
{}_{\mathrm{LC}}D^{\alpha}_{+}f(x) = {}_{\mathrm{L}}I^{1-\alpha}_{+}\frac{d}{dx}f(x) \tag{5.36}
$$

$$
= \frac{1}{\Gamma(1-\alpha)}\int_{-\infty}^{x}(x-\xi)^{-\alpha}\frac{df(\xi)}{d\xi}d\xi \tag{5.37}
$$

$$
= \frac{1}{\Gamma(1-\alpha)}\int_{-\infty}^{x}(x-\xi)^{-\alpha}f'(\xi)d\xi \tag{5.38}
$$

$$
{}_{\mathrm{LC}}D^{\alpha}_{-}f(x) = {}_{\mathrm{L}}I^{1-\alpha}_{-}\frac{d}{dx}f(x) \tag{5.39}
$$

$$
= \frac{1}{\Gamma(1-\alpha)}\int_{x}^{+\infty}(\xi-x)^{-\alpha}\frac{df(\xi)}{d\xi}d\xi \tag{5.40}
$$

$$
= \frac{1}{\Gamma(1-\alpha)}\int_{x}^{+\infty}(\xi-x)^{-\alpha}f'(\xi)d\xi \tag{5.41}
$$

This definition of a fractional derivative has no special name nor did we find any mention in the mathematical literature. At first it could be assumed, that both derivatives (5.28) and (5.36) lead to similar results. But with the special choice $f(x)$ = const it follows immediately that the Liouville fractional derivative definition (5.28) leads to a divergent result, while with (5.36) the result vanishes.

As a general remark, the conditions for analytic functions which lead to a convergent result for the Liouville fractional derivative are much more restrictive than for the Liouville fractional derivative with inverted operator sequence.

As a convergence condition for the Liouville derivative definition we obtain for an analytic function

$$\lim_{x \to \pm\infty} f(x) = \mathrm{O}(|x|^{-\alpha-\epsilon}), \qquad \epsilon > 0 \tag{5.42}$$

while for the Liouville derivative with inverted operator sequence the same behaviour is required only for the derivative.

The idea to define a fractional derivative with an inverted operator sequence was first proposed by Caputo and was used for a reformulation of the Riemann fractional derivative. Hence we will call the fractional derivative according to Liouville with inverted operator sequence as Liouville–Caputo derivative.

As long as both integrals (5.28) and (5.36) are convergent for a given function, the Liouville and the Liouville–Caputo derivative definition lead to similar results.

Later in Chapter 8 we will demonstrate, that both definitions will lead to different results when extended to describe general non-local operators.

In Table 5.4 some functions and the corresponding Liouville–Caputo derivative are listed.

A comparison with Table 5.2 demonstrates, that both definitions lead to similar results, as long as the integrals (5.28) and (5.36) converge.

On the other hand, the error function $\mathrm{erf}(x)$ is a nice example for the different convergence requirements of the Liouville and Liouville–Caputo fractional derivative definition.

In the sense of Liouville the fractional derivative integral (5.28) applied to the error function is divergent and therefore the error function is not differentiable. For the Liouville-Caputo definition of a fractional derivative the integral (5.36) converges and therefore the error function is differentiable.

Table 5.4. Some special functions and the corresponding Liouville–Caputo derivative ${}_{\text{LC}}D_+$

| $f(x)$ | $\frac{d^\alpha}{dx^\alpha} f(x)$ |
|---|---|
| $e^{kx}$ | $k^\alpha e^{kx}, \quad k \geq 0$ |
| $\sin(kx)$ | $k^\alpha \sin(kx + \frac{\pi}{2}\alpha)$ |
| $\cos(kx)$ | $k^\alpha \cos(kx + \frac{\pi}{2}\alpha)$ |
| $e^{-kx^2}$ | $\frac{k^{\frac{\alpha}{2}}}{\Gamma(1-\alpha)} \Big(\Gamma(1-\frac{\alpha}{2})_1F_1(\frac{1}{2}+\frac{\alpha}{2};\frac{1}{2};-kx^2)$ $-\sqrt{k}\alpha x\Gamma(\frac{1}{2}-\frac{\alpha}{2})_1F_1(1+\frac{\alpha}{2};\frac{3}{2};-kx^2)\Big)$ |
| $\text{erf}(kx)$ | $-\frac{k^\alpha}{\pi^{3/2}}\sin(\pi\alpha)\,(kx\Gamma(-\frac{\alpha}{2})\Gamma(1+\alpha)_1F_1(\frac{1}{2}+\frac{\alpha}{2};\frac{3}{2};-k^2x^2)$ $-\Gamma(\frac{1}{2}-\frac{\alpha}{2})\Gamma(\alpha)_1F_1(\frac{\alpha}{2};\frac{1}{2};-k^2x^2))$ |
| ${}_pF_q(\{a_i\};\{b_j\};kx)$ | $k^\alpha \prod_{i=1}^{p} \frac{\Gamma(a_i+\alpha)}{\Gamma(a_i)} \prod_{j=1}^{q} \frac{\Gamma(b_j)}{\Gamma(b_j+\alpha)}$ ${}_pF_q(\{a_i+\alpha\};\{b_j+\alpha\};kx)$ |

### 5.3.4 *The Riemann fractional derivative with inverted operator sequence: the Caputo fractional derivative*

The inverted operator sequence (5.26) using the Riemann fractional integral definition (5.17), (5.18) leads to:

$$
\begin{aligned}
{}_{\text{C}}D_+^\alpha f(x) &= {}_{\text{R}}I_+^{1-\alpha}\frac{d}{dx}f(x) && (5.43)\\
&= \frac{1}{\Gamma(1-\alpha)}\int_0^x (x-\xi)^{-\alpha}\frac{df(\xi)}{d\xi}d\xi && (5.44)\\
&= \frac{1}{\Gamma(1-\alpha)}\int_0^x (x-\xi)^{-\alpha}f'(\xi)d\xi && (5.45)\\
{}_{\text{C}}D_-^\alpha f(x) &= {}_{\text{R}}I_-^{1-\alpha}\frac{d}{dx}f(x) && (5.46)\\
&= \frac{1}{\Gamma(1-\alpha)}\int_x^0 (\xi-x)^{-\alpha}\frac{df(\xi)}{d\xi}d\xi && (5.47)\\
&= \frac{1}{\Gamma(1-\alpha)}\int_x^0 (\xi-x)^{-\alpha}f'(\xi)d\xi && (5.48)
\end{aligned}
$$

This definition is known as the Caputo fractional derivative. The difference between Riemann and Caputo fractional derivative may be demonstrated using $f(x) = \text{const}$.

For the Riemann fractional derivative it follows

$$ {}_{\text{R}}D_+^\alpha \text{const} = \frac{\text{const}}{\Gamma(1-\alpha)}x^{-\alpha} \tag{5.49} $$

while applying the Caputo derivative we obtain

$$ {}_{\text{C}}D_+^\alpha \text{const} = 0 \tag{5.50} $$

Table 5.5. Some special functions and their Caputo derivative ${}_C D$

| $f(x)$ | $\frac{d^\alpha}{dx^\alpha} f(x)$ | |
|---|---|---|
| $e^{kx}$ | $\mathrm{sign}(x)(\mathrm{sign}(x)k)^\alpha e^{kx} \left(1 - \frac{\Gamma(1-\alpha,kx)}{\Gamma(1-\alpha)}\right)$ | |
| $x^p$ | $\frac{\Gamma(1+p)}{\Gamma(1+p-\alpha)} \mathrm{sign}(x)\|x\|^{-\alpha} x^p,$ | $p > -1$ |
| | $0,$ | $p = 0$ |
| $\sin(kx)$ | $\frac{k \mathrm{sign}(x)\|x\|^{-\alpha} x}{\Gamma(1-\alpha/2)\Gamma(3/2-\alpha/2)\Gamma(2-\alpha)} \, {}_1F_2(1; 1-\frac{\alpha}{2}, \frac{3}{2}-\frac{\alpha}{2}; -\frac{1}{4}k^2x^2)$ | |
| $\cos(kx)$ | $\frac{-k^2 \mathrm{sign}(x)\|x\|^{-\alpha} x^2}{\Gamma(3/2-\alpha/2)\Gamma(2-\alpha/2)\Gamma(3-\alpha)} \, {}_1F_2(1; \frac{3}{2}-\frac{\alpha}{2}, 2-\frac{\alpha}{2}; -\frac{1}{4}k^2x^2)$ | |
| $e^{-kx^2}$ | $\frac{-2kx^{2-\alpha}}{\Gamma(3-\alpha)} \, {}_2F_2(1, \frac{3}{2}; \frac{3}{2}-\frac{\alpha}{2}, 2-\frac{\alpha}{2}; -kx^2)$ | |
| $\mathrm{erf}(kx)$ | $\frac{2k \mathrm{sign}(x)\|x\|^{-\alpha} x}{\Gamma(2-a)\sqrt{\pi}} \, {}_2F_2(\frac{1}{2}, 1; 1-\frac{\alpha}{2}, \frac{3}{2}-\frac{\alpha}{2}; -k^2x^2)$ | |
| ${}_pF_q(\{a_i\}; \{b_j\}; kx)$ | $k \mathrm{sign}(x)\|x\|^{-\alpha} x \prod_{i=1}^p a_i \prod_{j=1}^q \Gamma(b_j)$ $\times \, {}_{p+1}F_{q+1}(\{1, 1+a_i\}; \{b_j, 2-\alpha\}; kx)$ | |

In Table 5.5 we have listed some special functions and the corresponding Caputo derivative.

Therefore in this section we have presented all standard definitions of a fractional derivative, which satisfy the condition

$$\lim_{\alpha \to 1} D^\alpha f(x) = \frac{df(x)}{dx} \tag{5.51}$$

## 5.4 Fractional derivative of second order

For the case $0 \leq \alpha \leq 1$ we have presented fractional derivative definitions. The four different definitions namely the Liouville, Riemann, Liouville–Caputo and Caputo derivatives are direct generalizations of the standard first order derivative to fractional values $\alpha$.

Derivatives of higher order may be generated by multiple application of a first order derivative:

$$\frac{d^n}{dx^n} = \frac{d}{dx}\frac{d}{dx} \cdots \frac{d}{dx} \qquad \text{n-fold} \tag{5.52}$$

This strategy is valid for fractional derivatives of higher order too:

$$D_\pm^\alpha f(x) = \begin{cases} \pm (D^m \, {}_aI_\pm^{m-a}) f(x), & m-1 < \alpha \leq m, \quad m \text{ odd}, \\ (D^m \, {}_aI_\pm^{m-a}) f(x), & m-1 < \alpha \leq m, \quad m \text{ even} \end{cases} \tag{5.53}$$

An alternative approach is based on the idea, to extend a derivative of higher order directly. In this section we will investigate the behaviour of fractional derivatives with the property

$$\lim_{\alpha \to 2} = \frac{d^2}{dx^2} \tag{5.54}$$

which of course not necessarily implies the similarity of the first derivative and the fractional derivative for $\alpha = 1$. There are well known strategies for a numerical solution of an ordinary differential equation which in a similar way approximate the second derivative directly (e.g. the Störmer-method).

### 5.4.1 *The Riesz fractional derivative*

For a fractional derivative definition which fulfils condition (5.54) the most prominent representative is the Riesz definition of a fractional derivative [Riesz(1949)].

Starting point is a linear combination of both fractional Liouville integrals (5.15) and (5.16)

$$ {}_{\text{RZ}}I^{\alpha} = \frac{{}_{\text{L}}I_{+}^{\alpha} + {}_{\text{L}}I_{-}^{\alpha}}{2\cos(\pi\alpha/2)} \tag{5.55}$$

$$ = \frac{1}{2\Gamma(\alpha)\cos(\pi\alpha/2)} \int_{-\infty}^{+\infty} |(x-\xi)|^{\alpha-1} f(\xi) d\xi, \tag{5.56}$$

$$ \alpha > 0, \, \alpha \neq 1, 3, 5, ...$$

which defines the so called Riesz potential or the fractional Riesz integral respectively.

In order to derive the explicit form of the Riesz fractional derivative we first present the left- and right handed Liouville derivative (5.15) and (5.16) in an alternative form:

$$ {}_{\text{L}}D_{+}^{\alpha} f(x) = \frac{\alpha}{\Gamma(1-\alpha)} \int_{0}^{\infty} \frac{f(x) - f(x-\xi)}{\xi^{\alpha+1}} d\xi \tag{5.57}$$

$$ {}_{\text{L}}D_{-}^{\alpha} f(x) = \frac{\alpha}{\Gamma(1-\alpha)} \int_{0}^{\infty} \frac{f(x) - f(x+\xi)}{\xi^{\alpha+1}} d\xi \tag{5.58}$$

which follows from (5.28) by partial integration:

$$ {}_{\text{L}}D_{+}^{\alpha} f(x) = {}_{\text{L}}I_{+}^{-\alpha} f(x) = +\frac{\partial}{\partial x} {}_{\text{L}}I_{+}^{1-\alpha} f(x) \tag{5.59}$$

$$ = \frac{1}{\Gamma(1-\alpha)} \frac{\partial}{\partial x} \int_{-\infty}^{x} (x-\xi)^{-\alpha} f(\xi) d\xi \tag{5.60}$$

$$ = \frac{1}{\Gamma(1-\alpha)} \frac{\partial}{\partial x} \int_{0}^{\infty} \xi^{-\alpha} f(x-\xi) d\xi \tag{5.61}$$

$$ = \frac{1}{\Gamma(1-\alpha)} \int_{0}^{\infty} \xi^{-\alpha} \left(-\frac{\partial}{\partial \xi} f(x-\xi)\right) d\xi \tag{5.62}$$

$$ = \frac{\alpha}{\Gamma(1-\alpha)} \left( \int_{0}^{\infty} \frac{f(x)}{\xi^{\alpha+1}} d\xi - \int_{0}^{\infty} \frac{f(x-\xi)}{\xi^{\alpha+1}} d\xi \right) \tag{5.63}$$

Table 5.6. Special functions and their Riesz fractional derivative ${}_{\mathrm{RZ}}D_{+}$

| $f(x)$ | $\frac{d^\alpha}{dx^\alpha} f(x)$ |
|---|---|
| $e^{ikx}$ | $-\lvert k\rvert^\alpha e^{ikx}$ |
| $\sin(kx)$ | $-\lvert k\rvert^\alpha \sin(kx)$ |
| $\cos(kx)$ | $-\lvert k\rvert^\alpha \cos(kx)$ |
| $e^{-kx^2}$ | $\frac{k^{\alpha/2}}{\pi} \sin(\frac{\pi\alpha}{2})\Gamma(-\alpha/2)\Gamma(1+\alpha){}_1F_1(\frac{1+\alpha}{2}; \frac{1}{2}; -kx^2), \quad k \geq 0$ |
| ${}_pF_q(\{a_i\}; \{b_j\}; kx)$ | divergent |

and similarly for (5.29)

$$
\begin{aligned}
{}_{\mathrm{L}}D_{-}^{\alpha} f(x) = {}_{\mathrm{L}}I_{-}^{-\alpha} f(x) &= -\frac{\partial}{\partial x} {}_{\mathrm{L}}I_{-}^{1-\alpha} f(x) && (5.64)\\
&= -\frac{1}{\Gamma(1-\alpha)} \frac{\partial}{\partial x} \int_x^\infty (\xi - x)^{-\alpha} f(\xi) d\xi && (5.65)\\
&= -\frac{1}{\Gamma(1-\alpha)} \frac{\partial}{\partial x} \int_0^\infty \xi^{-\alpha} f(\xi + x) d\xi && (5.66)\\
&= -\frac{1}{\Gamma(1-\alpha)} \int_0^\infty \xi^{-\alpha} \left(\frac{\partial}{\partial \xi} f(\xi + x)\right) d\xi && (5.67)\\
&= \frac{\alpha}{\Gamma(1-\alpha)} \left(\int_0^\infty \frac{f(x)}{\xi^{\alpha+1}} d\xi - \int_0^\infty \frac{f(\xi + x)}{\xi^{\alpha+1}} d\xi\right) && (5.68)
\end{aligned}
$$

Applying the reflection formula (2.11) we obtain for the factor

$$
\frac{\alpha}{\Gamma(1-\alpha)} = \Gamma(1+\alpha) \frac{\sin(\pi\alpha)}{\pi} \tag{5.69}
$$

With the definition of the fractional derivative according to Riesz

$$
{}_{\mathrm{RZ}}D^\alpha = -\frac{{}_{\mathrm{L}}D_{-}^{\alpha} + {}_{\mathrm{L}}D_{+}^{\alpha}}{2\cos(\pi\alpha/2)} \tag{5.70}
$$

we explicitly obtain

$$
{}_{\mathrm{RZ}}D^\alpha = \Gamma(1+\alpha) \frac{\sin(\pi\alpha/2)}{\pi} \int_0^\infty \frac{f(x+\xi) - 2f(x) + f(x-\xi)}{\xi^{\alpha+1}} d\xi, \; 0 < \alpha < 2 \tag{5.71}
$$

In Table 5.6 we have listed some examples for functions and their corresponding Riesz derivative. For $\alpha = 2$ the given fractional derivatives coincide with the standard second order derivative for the given functions.

An important property of the fractional Riesz derivative is the invariance of the scalar product

$$
\int_{-\infty}^\infty \left({}_{\mathrm{RZ}}D^{\alpha *} f^*(x)\right) g(x) dx = \int_{-\infty}^\infty f(x)^* \left({}_{\mathrm{RZ}}D^\alpha g(x)\right) dx \tag{5.72}
$$

where * designates complex conjugation.

### 5.4.2 *The Feller fractional derivative*

A possible generalization for the Riesz fractional derivative was proposed by Feller [Feller(1952)]. He suggested a general superposition of both fractional Liouville integrals:

$$ {}_{\mathrm{F}}I^{\alpha}_{\theta} = c_{-}(\theta,\alpha)_{\mathrm{L}}I^{\alpha}_{+} + c_{+}(\theta,\alpha)_{\mathrm{L}}I^{\alpha}_{-} \tag{5.73}$$

introducing a free parameter $0 < \theta < 1$ which is a measure for the influence of both components:

$$ c_{-}(\theta,\alpha) = \frac{\sin((\alpha-\theta)\pi/2)}{\sin(\pi\theta)} \tag{5.74}$$

$$ c_{+}(\theta,\alpha) = \frac{\sin((\alpha+\theta)\pi/2)}{\sin(\pi\theta)} \tag{5.75}$$

The fractional Feller derivative is then given as

$$ {}_{\mathrm{F}}D^{\alpha}_{\theta} = -\left(c_{+}(\theta,\alpha)_{\mathrm{L}}D^{\alpha}_{+} + c_{-}(\theta,\alpha)_{\mathrm{L}}D^{\alpha}_{-}\right) \tag{5.76}$$

For the special case $\theta = 0$ we obtain

$$ c_{-}(\theta=0,\alpha) = c_{+}(\theta=0,\alpha) = \frac{1}{2\cos(\alpha\pi/2)} \tag{5.77}$$

which exactly corresponds to the definition of the Riesz derivative (5.71).

There is an additional special case for $\theta = 1$

$$ c_{-}(\theta=1,\alpha) = -c_{+}(\theta=1,\alpha) = \frac{1}{2\sin(\alpha\pi/2)} \tag{5.78}$$

which leads to a very simple realization of a fractional derivative:

$$ {}_{\mathrm{F}}D^{\alpha}_{1}f(x) = \frac{{}_{\mathrm{L}}D^{\alpha}_{+} - {}_{\mathrm{L}}D^{\alpha}_{-}}{2\sin(\alpha\pi/2)}\phi(x) \tag{5.79}$$

$$ = \Gamma(1+\alpha)\frac{\cos(\alpha\pi/2)}{\pi}\int_0^{\infty}\frac{f(x+\xi)-f(x-\xi)}{\xi^{\alpha+1}}d\xi \tag{5.80}$$

$$ 0 \leq \alpha < 1 $$

Since this derivative is a symmetric combination of the left- and right handed Liouville derivative (5.28), (5.29) it may be interpreted as a regularized Liouville derivative.

The Feller derivative therefore may be written as a linear combination of ${}_{\mathrm{F}}D^{\alpha}_{1}$ and ${}_{\mathrm{RZ}}D^{\alpha}$:

$$ {}_{\mathrm{F}}D^{\alpha}_{\theta} = A_1(\theta,\alpha)(D^{\alpha}_{+} - D^{\alpha}_{-}) + A_2(\theta,\alpha)(D^{\alpha}_{+} + D^{\alpha}_{-}) \tag{5.81}$$

Table 5.7. Some functions and the corresponding Feller derivative for $\theta = 1$, or ${}_{\mathrm{F}}D_1^\alpha$

| $f(x)$ | $\frac{d^\alpha}{dx^\alpha} f(x)$ | |
|---|---|---|
| $\sin(kx)$ | $\mathrm{sign}(k)\lvert k\rvert^\alpha \cos(kx)$ | |
| $\cos(kx)$ | $-\mathrm{sign}(k)\lvert k\rvert^\alpha \sin(kx)$ | |
| $e^{-kx^2}$ | $-\frac{k^{\frac{1+\alpha}{2}}}{\pi} x \cos(\pi\alpha/2)\Gamma(\frac{1-\alpha}{2})\Gamma(1+\alpha){}_1F_1(\frac{2+\alpha}{2}; \frac{3}{2}; -kx^2)$, | $k \geq 0$ |
| ${}_pF_q(\{a_i\}; \{b_j\}; kx)$ | divergent | |

with coefficients

$$A_1(\theta, \alpha) = -\frac{1}{2}\left(c_+(\theta, \alpha) - c_-(\theta, \alpha)\right) = -\frac{\sin(\theta\pi/2)}{2\sin(\alpha\pi/2)} \tag{5.82}$$

$$A_2(\theta, \alpha) = -\frac{1}{2}\left(c_+(\theta, \alpha) + c_-(\theta, \alpha)\right) = -\frac{\cos(\theta\pi/2)}{2\cos(\alpha\pi/2)} \tag{5.83}$$

Hence we obtain:

$${}_{\mathrm{F}}D_\theta^\alpha = \sin(\theta\pi/2)\,{}_{\mathrm{F}}D_1^\alpha + \cos(\theta\pi/2)\,{}_{\mathrm{RZ}}D^\alpha \tag{5.84}$$

or more accurately

$${}_{\mathrm{F}}D_\theta^\alpha = \begin{cases} \sin(\theta\pi/2)\,{}_{\mathrm{F}}D_1^\alpha + \cos(\theta\pi/2)\,{}_{\mathrm{RZ}}D^\alpha & 0 < \alpha < 1 \\ \sin(\theta\pi/2)\frac{d}{dx}\,{}_{\mathrm{F}}D_1^{\alpha-1} + \cos(\theta\pi/2)\,{}_{\mathrm{RZ}}D^\alpha & 1 \leq \alpha < 2 \\ \sin(\theta\pi/2)\frac{d^2}{dx^2}\,{}_{\mathrm{F}}D_1^{\alpha-2} + \cos(\theta\pi/2)\frac{d^2}{dx^2}\,{}_{\mathrm{RZ}}D^{\alpha-2} & 2 \leq \alpha < 3 \\ \vdots & \end{cases} \tag{5.85}$$

Having derived the Feller derivative in this form, the parameter $\theta$ may be understood as a rotation parameter. In addition we can use this form for an extension of this concept to fractional derivatives of higher orders.

## 5.5 Fractional derivatives of higher orders

Second order differential equations play a central role in physics, e.g. the description of oscillations, waves or the whole area of quantum mechanics. But there are subjects and problems which are better described using higher order derivatives.

A remarkable example for a non linear third order differential equation is the Korteweg–de Vries equation [Korteweg(1895)]:

$$\left(\frac{\partial}{\partial t} + \frac{\partial^3}{\partial x^3} - 6u\frac{\partial}{\partial x}\right) u(t, x) = 0 \tag{5.86}$$

The stationary solutions of this equation play an important role in the description of the behaviour of solitons.

Typical applications of fourth order differential equations are given in elasticity theory e.g. for elastic deflection under load of a rod we have to solve:

$$\left(\frac{\partial^4}{\partial x^4} + N\frac{\partial^2}{\partial x^2}\right) u(x) = q(x) \tag{5.87}$$

An example for a fifth order differential equation is the extended Korteweg–de Vries equation [Marchant(1990)]:

$$\left(\frac{\partial}{\partial t} + u\frac{\partial}{\partial x} - \frac{\partial^5}{\partial x^5}\right) u(t,x) = 0 \tag{5.88}$$

We have demonstrated that a large number of physical processes may be adequately described using differential equations of higher order. Since we interpret higher order derivatives as entities, which should directly correspond to a fractional extended pendant, we should present higher order fractional derivatives too.

Our derivation of the Riesz and Feller fractional derivative presented in the previous section provides a direct hint to the construction principle of higher fractional derivatives. Especially Feller's fractional first derivative (5.79) and Riesz's fractional second derivative (5.71) may be easily interpreted as integrals over differential approximations of the standard first and second order derivative. Using the basic properties of the central difference operators $\delta_{\frac{1}{2}}$ and $\delta_1$

$$\delta_{\frac{1}{2}}\phi(x) = \phi\left(x + \frac{1}{2}\xi\right) - \phi\left(x - \frac{1}{2}\xi\right) \tag{5.89}$$

$$\delta_1\phi(x) = \frac{1}{2}(\phi(x+\xi) - \phi(x-\xi)) \tag{5.90}$$

we define the central difference operator $\mathfrak{D}^k$ of order $k$

$$\mathfrak{D}^k f(x) = \begin{cases} \delta_{\frac{1}{2}}^k f(x) & \text{k even} \\ \delta_1 \delta_{\frac{1}{2}}^{k-1} f(x) & \text{k odd} \end{cases} \tag{5.91}$$

or explicitly, using (5.89) and (5.90):

$$\mathfrak{D}^k\phi(x) = \sum_{n=0}^{2[(k+1)/2]} a_n^k \phi(x - ([(k+1)/2] - n)\xi) \tag{5.92}$$

with the summation coefficients

$$a_n^k = (-1)^n \begin{cases} \binom{k}{n} & \text{k even} \\ \frac{1}{2}\left\{\binom{k-1}{n} - \binom{k-1}{n-2}\right\} & \text{k odd} \end{cases} \tag{5.93}$$

Table 5.8. Some special functions and the corresponding fractional derivatives according to ${}_3D^\alpha$. In the limit $\alpha \to 3$ the fractional derivatives coincide with the standard third order derivative

| $f(x)$ | $\frac{d^\alpha}{dx^\alpha} f(x)$ |
|---|---|
| $e^{ikx}$ | $-i\text{sign}(k)\|k\|^\alpha e^{ikx}$ |
| $\sin(kx)$ | $-\|k\|^\alpha \cos(kx)$ |
| $\cos(kx)$ | $-\|k\|^\alpha \sin(kx)$ |
| $e^{-kx^2}$ | $\frac{2^\alpha k^{\frac{1+\alpha}{2}} \sqrt{\pi}\alpha}{\sin(\pi\alpha/2)\Gamma(\frac{1-\alpha}{2})} x \, {}_1F_1(\frac{2+\alpha}{2}; \frac{3}{2}; -kx^2), \quad k \geq 0$ |
| ${}_pF_q(\{a_i\}; \{b_j\}; kx)$ | divergent |

The renormalized fractional derivative is then given as:

$$
{}_kD^\alpha \phi(x) = \frac{1}{N_k} \int_0^\infty \frac{d\xi}{\xi^{\alpha+1}} \mathfrak{D}^k \phi(x) \tag{5.94}
$$

and the normalization factor results as:

$$
\begin{aligned}
N_k = \frac{\Gamma(1+\alpha)}{\pi} & \left( 2 \sum_{n=0}^{[(k+1)/2]} a_n^k (k-n-1)^\alpha \right)^{-1} \\
& \times \begin{cases} (-1)^{\frac{k+2}{2}} \sin(\pi\alpha/2) & \text{k even} \\ (-1)^{\frac{k+1}{2}} \cos(\pi\alpha/2) & \text{k odd} \end{cases}
\end{aligned} \tag{5.95}
$$

With (5.94) based on the Liouville definition of the fractional derivative (5.28) we therefore have given all fractional derivatives, which extend the ordinary derivative of order $k$:

$$
\lim_{\alpha \to k} {}_kD^\alpha = \frac{d^k}{dx^k} \tag{5.96}
$$

In addition, for these derivatives the invariance of the scalar product follows:

$$
\int_{-\infty}^{\infty} \left({}_kD^{\alpha *} f^*(x)\right) g(x) dx = (\pm)^k \int_{-\infty}^{\infty} f(x)^* \left({}_kD^\alpha g(x)\right) dx \tag{5.97}
$$

Table 5.9. Some special functions and the corresponding fractional derivatives according to ${}_4D^\alpha$. In the limit $\alpha \to 4$ the fractional derivatives coincide with the standard fourth order derivative

| $f(x)$ | $\frac{d^\alpha}{dx^\alpha} f(x)$ |
|---|---|
| $e^{ikx}$ | $\|k\|^\alpha e^{ikx}$ |
| $\sin(kx)$ | $\|k\|^\alpha \sin(kx)$ |
| $\cos(kx)$ | $\|k\|^\alpha \cos(kx)$ |
| $e^{-kx^2}$ | $\frac{k^{\frac{\alpha}{2}}\Gamma(-\alpha/2)}{2\cos(\pi\alpha/2)\Gamma(-\alpha)}\, {}_1F_1(\frac{1+\alpha}{2}; \frac{1}{2}; -kx^2), \quad k \geq 0$ |
| ${}_pF_q(\{a_i\}; \{b_j\}; kx)$ | divergent |

The first four fractional derivative definitions according (5.94) follow as:

$$
\begin{aligned}
{}_1D^\alpha f(x) = & \Gamma(1+\alpha)\frac{\cos(\alpha\pi/2)}{\pi} \\
& \times \int_0^\infty \frac{f(x+\xi)-f(x-\xi)}{\xi^{\alpha+1}} d\xi \\
& \qquad 0 \leq \alpha < 1
\end{aligned}
\tag{5.98}
$$

$$
\begin{aligned}
{}_2D^\alpha f(x) = & \Gamma(1+\alpha)\frac{\sin(\alpha\pi/2)}{\pi} \\
& \times \int_0^\infty \frac{f(x+\xi)-2f(x)+f(x-\xi)}{\xi^{\alpha+1}} d\xi \\
& \qquad 0 \leq \alpha < 2
\end{aligned}
\tag{5.99}
$$

$$
\begin{aligned}
{}_3D^\alpha f(x) = & \Gamma(1+\alpha)\frac{\cos(\alpha\pi/2)}{\pi}\frac{1}{2^\alpha - 2} \\
& \times \int_0^\infty \frac{-f(x+2\xi)+2f(x+\xi)-2f(x-\xi)+f(x-2\xi)}{\xi^{\alpha+1}} d\xi \\
& \qquad 0 \leq \alpha < 3
\end{aligned}
\tag{5.100}
$$

$$
\begin{aligned}
{}_4D^\alpha f(x) = & \Gamma(1+\alpha)\frac{\sin(lpha\pi/2)}{\pi}\frac{1}{2^\alpha - 4} \\
& \times \int_0^\infty \frac{-f(x+2\xi)+4f(x+\xi)-6f(x)+4f(x-\xi)-f(x-2\xi)}{\xi^{\alpha+1}} d\xi \\
& \qquad 0 \leq \alpha < 4
\end{aligned}
\tag{5.101}
$$

$$
\tag{5.102}
$$

These definitions are valid for $0 \leq \alpha < k$. Setting $\alpha > k$

$$
{}_kD^\alpha = \frac{d^{nk}}{dx^{nk}} {}_kD^{\alpha - nk}, \qquad n \in \mathbb{N} \tag{5.103}
$$

and choosing $n$ so that $0 \leq \alpha - nk < k$ the definitions given are valid for all $\alpha > 0$.

In the same manner the Feller fractional derivative definition may be extended to fractional derivatives of higher order.

We introduce hyperspherical coordinates on the unit sphere on $\mathbb{R}^n$:

$$x_1 = \cos(\theta_{n-1}) \tag{5.104}$$

$$x_2 = \sin(\theta_{n-1})\cos(\theta_{n-2}) \tag{5.105}$$

$$\vdots$$

$$x_{n-1} = \sin(\theta_{n-1})\sin(\theta_{n-2})...\cos(\theta_1) \tag{5.106}$$

$$x_n = \sin(\theta_{n-1})\sin(\theta_{n-2})...\sin(\theta_1) \tag{5.107}$$

With these coordinates the Feller definition of a fractional derivative may be extended to

$$_{\text{F}}D^{\alpha}_{\{\theta_k\}} = \sum_{k=1}^{n} x_k \, {}_kD^{\alpha} \tag{5.108}$$

## 5.6 Geometric interpretation of the fractional integral

A geometric interpretation of a fractional integral has first been proposed by Podlubny [Podlubny(2002)].

We illustrate his idea using the left-handed fractional Riemann integral (5.17).

$$_{\text{R}}I^{\alpha}_{+}f(x) = \frac{1}{\Gamma(\alpha)} \int_0^x (x-\xi)^{\alpha-1} f(\xi) d\xi \tag{5.109}$$

Introducing an auxiliary function $g_x(\xi)$ we can write the integral as

$$_{\text{R}}I^{\alpha}_{+}f(x) = \frac{1}{\Gamma(\alpha)} \int_0^x f(\xi) dg_x(\xi) \tag{5.110}$$

with

$$g_x(\xi) = \frac{1}{\Gamma(1+\alpha)} (x^{\alpha} - (x-\xi)^{\alpha}) \tag{5.111}$$

For a given upper limit $x$ of the integral (5.109) we define a three dimensional coordinate system with coordinates $\{\xi, g, f\}$. In the $(\xi, g)$-plane or, as illustrated in Fig. 5.1 on the floor, we first draw the function $g_x(\xi)$ in the region $0 \leq \xi \leq x$. Along this path we draw $f(\xi)$, in this way we construct a fence (red colour). The upper boundary of the fence is a three dimensional curve $(\xi, g_x(\xi), f(\xi))$ in the interval $0 \leq \xi \leq x$. This fence may now be projected to two different planes: A first projection onto the

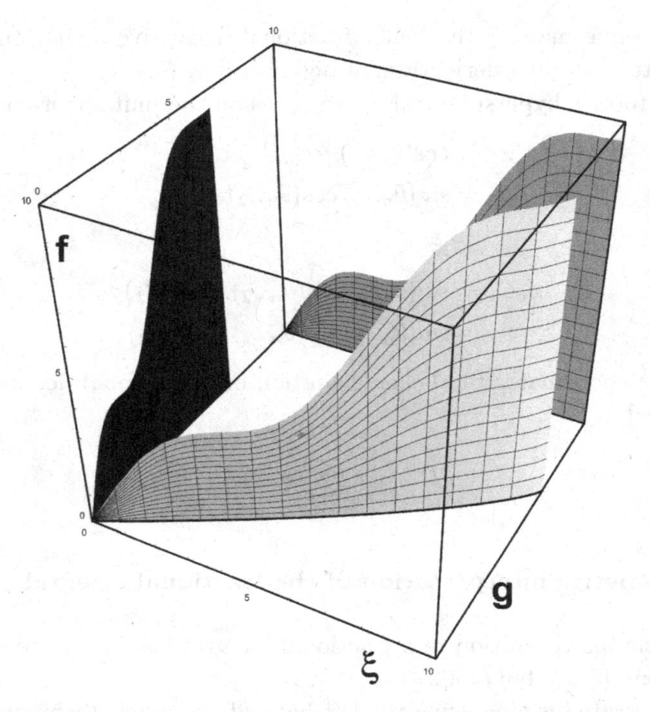


Fig. 5.1. The fence and its projections. The geometric interpretation of an integral is given in terms of a projected area of the light gray fence. At the back in dark gray is the standard integral. On the left plane the fractional integral is given as the black area from the fence projection.

$(\xi, f)$-plane generates the area below the curve $f(\xi)$, which corresponds to the standard integral

$$ {}_{\mathrm{R}}I_{+}^{1} f(x) = \int_{0}^{x} f(\xi) d\xi \tag{5.112} $$

This area is dark gray in Fig. 5.1.

A second projection onto the $(g, f)$-plane, which corresponds to the area of the integral (5.110) is coloured black in Fig. 5.1. Therefore the black area is a geometric representation for the fractional integral.

In the special case $\alpha = 1$ $g_x(\xi) = \xi$ and both projections are of similar form. Hence we have a smooth transition from a standard to a fractional integral.

A variation of the upper bound $x$ results in a change of the path $g_x(\xi)$ and consequently the shape of projections is changed. On the left side of

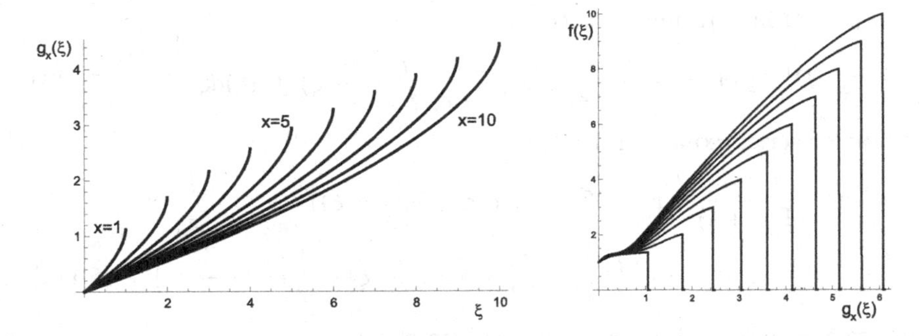


Fig. 5.2. The fence path and projection of the fractional integral. For $\alpha = 0.6$, on the left the function $g_x(\xi)$ projected onto the $(g, \xi)$-plane is plotted for different $x = 1, 2, ..., 10$, which corresponds to the view of the fence from above. On the right the corresponding projections onto the $(g, f)$-plane of the fractional integral are given for $f(\xi) = \xi + e^{-\xi^2}$.

Fig. 5.2 we have drawn different paths and on the right side the corresponding projections onto the $(g, f)$-plane are given, which directly correspond to the value of the fractional integral for increasing $x$.

The change of the projected area in shape and size with increasing $x$ gives a direct geometric picture of the change of the fractional integral as a function of $x$.

## 5.7 Low level fractionality

In this chapter so far we have presented several commonly used definitions of the fractional derivative parametrized via an arbitrary derivative parameter $\alpha$. It is interesting to investigate cases where $\alpha$ only minimally deviates from an integer value.

$$\alpha = n - \epsilon \tag{5.113}$$

For $\epsilon \ll 1$ the behaviour of the fractional derivative is called low level fractionality and was considered e.g. by Tarasov [Tarasov(2006)].

For demonstrative purposes we will investigate low level fractionality for the left handed Caputo derivative (5.43) for the simple case $\alpha = 1 - \epsilon$.

It follows

$$_{\mathrm{C}}D_{+}^{1-\epsilon} f(x) = \frac{1}{\Gamma(\epsilon)} \int_0^x (x - \xi)^{\epsilon - 1} f'(\xi) d\xi \tag{5.114}$$

and partial integration leads to:

$$_{\rm C}D_{+}^{1-\epsilon} f(x) = \frac{f'(0)x^\epsilon}{\Gamma(1+\epsilon)} + \frac{1}{\Gamma(1+\epsilon)} \int_0^x (x-\xi)^\epsilon f''(\xi) d\xi \tag{5.115}$$

A Taylor series expansion leads to:

$$\begin{aligned} \frac{1}{\Gamma(1+\epsilon)} \ (x-\xi)^\epsilon &= 1 + \epsilon\,(\gamma + \ln(x-\xi)) \\ &+\epsilon^2 \left( \frac{\gamma^2}{2} - \frac{\pi^2}{12} + \gamma \ln(x-\xi) + \frac{1}{2} \ln^2(x-\xi) \right) \end{aligned} \tag{5.116}$$

where $\gamma = 0.577...$ denotes Euler's constant.

In lowest order in $\epsilon$ therefore for the fractional derivative follows:

$$_{\rm C}D_{+}^{1-\epsilon} f(x) = f'(x) + \epsilon \left( f'(0) \ln(x) + \gamma f'(x) + \int_0^x f''(\xi) \ln(x-\xi) d\xi \right) \tag{5.117}$$

For $\epsilon \ll 1$ we obtain the result:

$$_{\rm C}D_{+}^{1-\epsilon} f(x) = f'(x) + \epsilon D_1^1 f(x) + ... \tag{5.118}$$

with

$$D_1^1 = f'(0) \ln(x) + \gamma f'(x) + \int_0^x f''(\xi) \ln(x-\xi) d\xi \tag{5.119}$$

in the limit $\lim \epsilon \to 0$ we obtain:

$$\lim_{\epsilon \to 0} {}_{\rm C}D_{+}^{1-\epsilon} = f'(x) \tag{5.120}$$

Hence we have demonstrated that there is a smooth transition between fractional, low level and ordinary derivatives.

As an example we obtain for $f(x) = \cos(x)$:

$$_{\rm C}D_{+}^{1-\epsilon} \cos(x) = -\sin(x) + \epsilon\,(\cos(x)\mathrm{Si}(x) - \sin(x)\mathrm{Ci}(x)) \tag{5.121}$$

where Ci and Si are the cosine- (2.22) and sine-integral (2.32).

## 5.8 Discussion

### 5.8.1 *Semigroup property of the fractional integral*

**Question:**
How can I prove the semigroup property of the fractional integral

$$ {}_aI^{\alpha}_{+}{}_aI^{\beta}_{+}f(x) = {}_aI^{\alpha+\beta}_{+}f(x), \qquad \alpha, \beta \geq 0 \tag{5.122}$$

where ${}_aI^{\alpha}_{+}$ is given by (5.13)

**Answer:**
To solve that problem, we will use the definition of the beta function:

$$\int_0^1 d\xi\, \xi^{\alpha-1}(1-\xi)^{\beta-1} = \frac{\Gamma(\alpha)\Gamma(\beta)}{\Gamma(\alpha+\beta)} \tag{5.123}$$

and the Dirichlet formula, which is given by [Whittaker(1965)]

$$\begin{aligned}\int_a^x d\xi\,(x-\xi)^{\alpha-1} &\int_a^{\xi} d\phi\,(\xi-\phi)^{\beta-1} g(\xi,\phi) \\ &= \int_a^x d\phi \int_{\phi}^x d\xi\,(x-\xi)^{\alpha-1}(\xi-\phi)^{\beta-1} g(\xi,\phi)\end{aligned} \tag{5.124}$$

which reduces for $g(\xi, \phi) = g(\xi)f(\phi)$ for the simplified case $g(\xi) = 1$ to:

$$\begin{aligned}\int_a^x d\xi\,(x-\xi)^{\alpha-1} &\int_a^{\xi} d\phi\,(\xi-\phi)^{\beta-1} f(\phi) \\ &= \int_a^x d\phi\, f(\phi) \int_{\phi}^x d\xi\,(x-\xi)^{\alpha-1}(\xi-\phi)^{\beta-1}\end{aligned} \tag{5.125}$$

Therefore we obtain

$$ {}_aI^{\alpha}_{+}{}_aI^{\beta}_{+}f(x) = \frac{1}{\Gamma(\alpha)}\frac{1}{\Gamma(\beta)} \int_a^x d\phi \int_a^{\phi} d\xi\,(x-\xi)^{\alpha-1}(\xi-\phi)^{\beta-1} f(\phi) \tag{5.126}$$

according to (5.125) this is equivalent to

$$ {}_aI^{\alpha}_{+}{}_aI^{\beta}_{+}f(x) = \frac{1}{\Gamma(\alpha)\Gamma(\beta)} \int_a^x d\phi\, f(\phi) \int_{\phi}^x d\xi\,(x-\xi)^{\alpha-1}(\xi-\phi)^{\beta-1} \tag{5.127}$$

The inner integral $k(x, \phi)$ in (5.127) may be interpreted as the kernel of the external convolution integral. With the substitution

$$u = \frac{\xi-\phi}{x-\phi} \tag{5.128}$$

which leads to $\xi = \phi + u(x-\phi)$ and $d\xi = (x-\phi)du$ we obtain

$$k(x,\phi) = \int_\phi^x d\xi\,(x-\xi)^{\alpha-1}(\xi-\phi)^{\beta-1} \tag{5.129}$$

$$= (x-\phi)^{\alpha+\beta-1}\int_0^1 du\,(1-u)^{\alpha-1}u^{\beta-1} \tag{5.130}$$

$$= (x-\phi)^{\alpha+\beta-1}\frac{\Gamma(\alpha)\Gamma(\beta)}{\Gamma(\alpha+\beta)} \tag{5.131}$$

In the last step we used the definition of the beta function. Inserting this result into (5.127) completes the proof.

$${}_aI_+^\alpha {}_aI_+^\beta f(x) = \frac{1}{\Gamma(\alpha+\beta)}\int_a^x d\phi\,(x-\xi)^{\alpha+\beta-1}f(\phi) \tag{5.132}$$

$$= {}_aI_+^{\alpha+\beta}f(x) \tag{5.133}$$

A similar procedure may be used for the right-handed integrals ${}_aI_-^\alpha$ too. Furthermore, since the derivation is valid for an arbitrary $a < x$, we have proven the validity of the semigroup property for the Liouville and the Riemann fractional integral definition as well.

## Chapter 6

# The Fractional Harmonic Oscillator

*There are only few problems which may be solved analytically in full generality. The solutions of the differential equation of the harmonic oscillator*

$$\left(m\frac{d^2}{dt^2}+k\right)x(t)=0 \tag{6.1}$$

*which are given by*

$$x(t)=c_1\cos(\omega t)+c_2\sin(\omega t) \tag{6.2}$$

*play a dominant role in theoretical physics. Furthermore this differential equation has a wide range of applications.*

*In the field of classical mechanics the equation (6.1) describes free oscillations and in general is a harmonic approximation for an arbitrary potential minimum. If appropriate boundary conditions are given this equation is a useful tool for describing of a vibrating string or a rectangular membrane as a spatial solution of a wave equation:*

$$\left(\frac{d^2}{dx^2}+\frac{d^2}{dy^2}+\frac{d^2}{dz^2}-\frac{1}{c^2}\frac{d^2}{dt^2}\right)\psi(x,y,z,t)=0 \tag{6.3}$$

*In quantum mechanics the same differential equation results as a special solution of the stationary Schrödinger equation*

$$-\frac{\hbar^2}{2m}\left(\frac{d^2}{dx^2}+\frac{d^2}{dy^2}+\frac{d^2}{dz^2}\right)\psi(x)=E\psi(x) \tag{6.4}$$

*for a particle confined to a bounded region of space.*

*Therefore it is of fundamental importance to study the fractional pendant of the harmonic oscillator and to discuss the specific properties of its solutions.*

*It will turn out, that the solutions for the fractional harmonic oscillator and the solutions for fractional friction forces given in the previous chapter show a similar behaviour. This similarity may be understood, if we consider the corresponding characteristic polynomials, which determine the solutions.*

## 6.1 The fractional harmonic oscillator

We define the following fractional differential equation

$$\left(m \frac{d^{2\alpha}}{dt^{2\alpha}} + k\right) x(t) = 0 \tag{6.5}$$

we are free to adjust the dimensions and meaning of the parameters $m$ and $k$ respectively or both $m, k$.

With the settings

$$\left(\frac{d^{2\alpha}}{dt^{2\alpha}} + \frac{k}{m}\right) x(t) = 0 \tag{6.6}$$

in units of the mks-system $[k/m]$ is given as $1/s^{2\alpha}$.

In the following we will present the solutions of this fractional differential equation for different types of the fractional derivative and will investigate the specific differences of the obtained solutions.

## 6.2 The harmonic oscillator according to Fourier

With the ansatz

$$x(t) = e^{\omega t} \tag{6.7}$$

the solution of the fractional oscillator is reduced to the examination of the zeroes of the polynomial

$$\omega^{2\alpha} + \frac{k}{m} = 0 \tag{6.8}$$

In the region $1/2 < \alpha \leq 3/2$ are exactly two complex conjugated solutions:

$$\omega_{1,2} = \left|\frac{k}{m}\right|^{\frac{1}{2\alpha}} \left(\cos\left(\frac{\pi}{2\alpha}\right) \pm i \sin\left(\frac{\pi}{2\alpha}\right)\right) \tag{6.9}$$

For $\alpha = 1$ we obtain two purely imaginary solutions, which corresponds to a free, undamped oscillation. If $\alpha$ is decreased, an increasing negative real part occurs which from a classical point of view may be interpreted as an increasing damping. For $\alpha > 1$ we obtain an increasing positive real part which corresponds to an increasing excitation.

A direct comparison with the classical solutions (4.40) of an harmonic oscillator including friction leads to:

$$-\frac{\gamma}{2m} = \left|\frac{k}{m}\right|^{\frac{1}{2\alpha}} \cos\left(\frac{\pi}{2\alpha}\right) \tag{6.10}$$

$$\approx \left|\frac{k}{m}\right|^{\frac{1}{2\alpha}} \frac{1}{2}\pi(\alpha - 1) \quad \text{for}\, \alpha \approx 1 \tag{6.11}$$

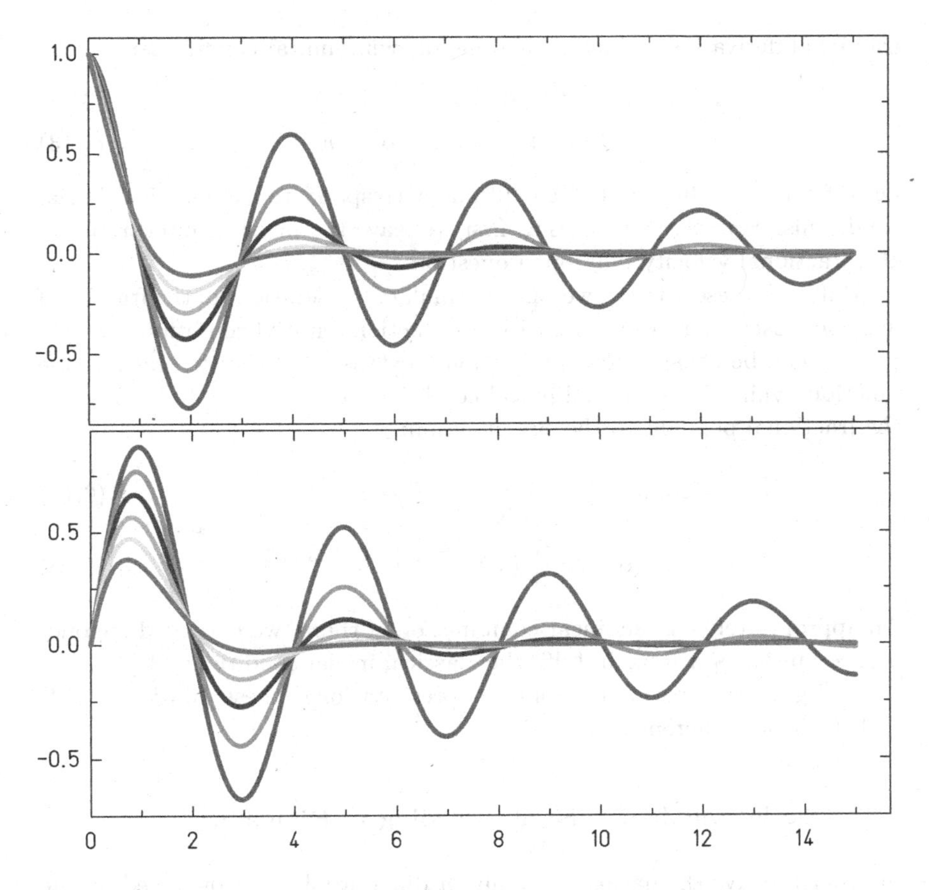


Fig. 6.1. Solutions ${}_F\cos(\alpha, t)$ and ${}_F\sin(\alpha, t)$ of the fractional harmonic oscillator using the Fourier derivative with $\alpha = 0.95, 0.90, 0.85, 0.80, 0.75, 0.70$.

Near $\alpha \approx \frac{1}{2}$ there is no oscillating contribution any more. The evolution in time for this system is dominated by an exponential decay. On the other hand $\alpha > 1$ corresponds to a negative classical friction coefficient. Consequently we obtain the surprising result, that the behaviour of the free solutions of the fractional harmonic oscillator under variation of the fractional derivative parameter $\alpha$ may be interpreted from a classical point of view as damping and excitation phenomena respectively.

In order to simplify a comparison for the solutions for different types of fractional derivatives we introduce the following initial conditions:

$$x(t = 0) = x_0 \tag{6.12}$$

$$D^\alpha x(t)|_{t=0} = \tilde{v}(t = 0) = \tilde{v}_0 \tag{6.13}$$

Only for $\alpha = 1$ these initial conditions correspond to the classical initial conditions. For all other cases at first we leave the physical interpretation of a fractional velocity $\tilde{v}$ an open question.

But with these settings we can reformulate the solutions of the fractional harmonic oscillator: a first set with the fractional initial conditions $x_0 = 1$, $\tilde{v}_0 = 0$ may be considered as a fractional extension of the standard cosine function, while the fractional initial conditions $x_0 = 0$, $\tilde{v}_0 = 1$ characterize the fractional pendant to the sine function:

$${}_{\mathrm{F}}\sin(\alpha, t) = \frac{1}{2i}\left(e^{\frac{1}{2}e^{\frac{i\pi}{2\alpha}}\pi t} - e^{\frac{1}{2}e^{-\frac{i\pi}{2\alpha}}\pi t}\right) \tag{6.14}$$

$${}_{\mathrm{F}}\cos(\alpha, t) = \frac{1}{2}\left(e^{\frac{1}{2}e^{\frac{i\pi}{2\alpha}}\pi t} + e^{\frac{1}{2}e^{-\frac{i\pi}{2\alpha}}\pi t}\right) \tag{6.15}$$

An appropriately chosen linear combination of these two extended trigonometric functions will again fulfil the classical initial conditions (4.3).

In Fig. 6.1 we present a graphical representation of these solutions (6.14) and (6.15) for different $\alpha$.

## 6.3 The harmonic oscillator according to Riemann

In order to solve the harmonic oscillator differential equation based on the Riemann definition of the fractional derivative we use a series expansion according to (3.20).

Since

$${}_{\mathrm{R}}D^\alpha t^{n\alpha} = \frac{\Gamma(1 + n\alpha)}{\Gamma(1 + (n-1)\alpha)} t^{(n-1)\alpha} \tag{6.16}$$

we obtain two different linearly independent solutions, which we call in analogy to the trigonometric functions ${}_{\mathrm{R}}\sin(\alpha, t)$ and ${}_{\mathrm{R}}\cos(\alpha, t)$,

$${}_{\mathrm{R}}\sin(\alpha, t) = t^{\alpha-1}\sum_{n=0}^{\infty}(-1)^n \frac{t^{(2n+1)\alpha}}{\Gamma((2n+2)\alpha)} \tag{6.17}$$

$${}_{\mathrm{R}}\cos(\alpha, t) = t^{\alpha-1}\sum_{n=0}^{\infty}(-1)^n \frac{t^{2n\alpha}}{\Gamma((2n+1)\alpha)} \tag{6.18}$$

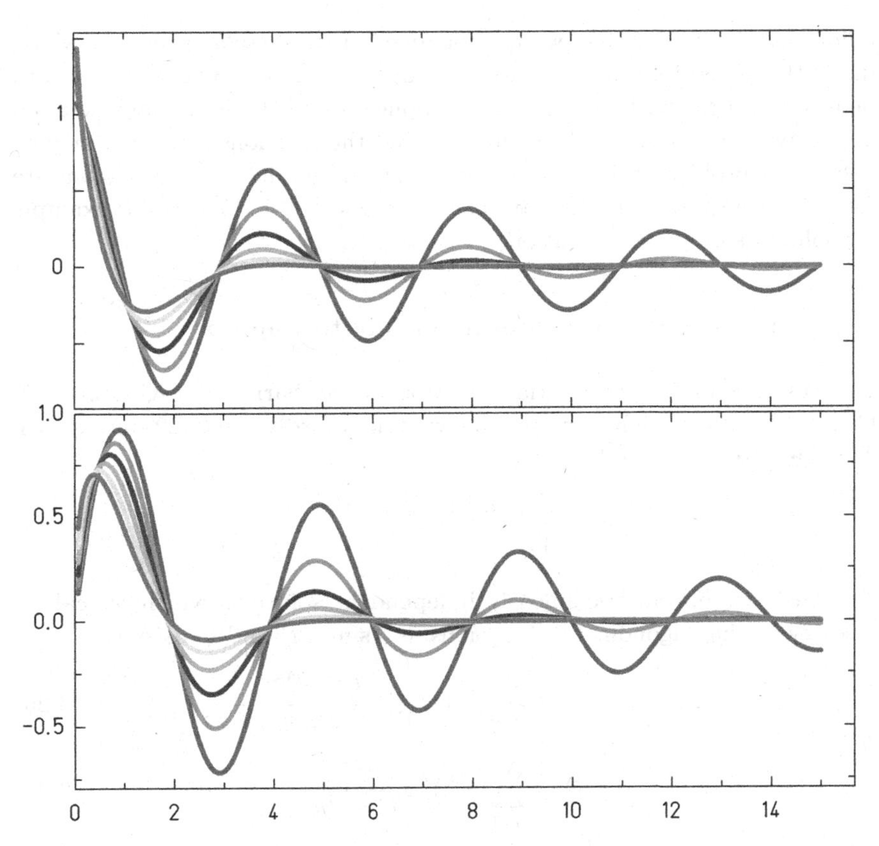


Fig. 6.2. Solutions ${}_{\mathrm{R}}\cos(\alpha, t)$ and ${}_{\mathrm{R}}\sin(\alpha, t)$ of the harmonic oscillator with Riemann derivative for $\alpha = 0.95, 0.90, 0.85, 0.80, 0.75, 0.70$.

with the property

$$ {}_{\mathrm{R}}D^{\alpha}\, {}_{\mathrm{R}}\sin(\alpha, \omega t) = \omega^{\alpha}\, {}_{\mathrm{R}}\cos(\alpha, \omega t) \tag{6.19} $$

$$ {}_{\mathrm{R}}D^{\alpha}\, {}_{\mathrm{R}}\cos(\alpha, \omega t) = -\omega^{\alpha}\, {}_{\mathrm{R}}\sin(\alpha, \omega t) \tag{6.20} $$

These functions are related to the Mittag-Leffler function (2.17):

$$ {}_{\mathrm{R}}\sin(\alpha, t) = t^{2\alpha-1} E(2\alpha, 2\alpha, -t^{2\alpha}) \tag{6.21} $$

$$ {}_{\mathrm{R}}\cos(\alpha, t) = t^{\alpha-1} E(2\alpha, \alpha, -t^{2\alpha}) \tag{6.22} $$

For $\alpha < 1$

$$ \lim_{t \to 0} {}_{\mathrm{R}}\cos(\alpha, \omega t) = \infty \tag{6.23} $$

holds, which means, that we cannot give a non singular solution, which fulfils the general initial conditions (6.12). At a first glance, this seems to be a serious drawback for a practical application of the Riemann fractional derivative. But we should keep in mind that the solutions presented may be helpful for problems which are not determined by initial conditions but are formulated in terms of boundary conditions, which is the case for example for solutions of a wave equation.

## 6.4 The harmonic oscillator according to Caputo

The solution of the differential equation of the harmonic oscillator with Caputo derivative definition may be written according to (3.22) as a series. The Caputo derivative for a power is given as

$$ {}_{\mathrm{C}}D^{\alpha} t^{n\alpha} = \begin{cases} \frac{\Gamma(1+n\alpha)}{\Gamma(1+(n-1)\alpha)} t^{(n-1)\alpha} & n > 0, \\ 0 & n = 0 \end{cases} \tag{6.24} $$

Therefore we obtain two linearly independent solutions which we call in analogy to the trigonometric functions as ${}_{\mathrm{C}}\sin(\alpha, t)$ and ${}_{\mathrm{C}}\cos(\alpha, t)$:

$$ {}_{\mathrm{C}}\sin(\alpha, t) = \sum_{n=0}^{\infty} (-1)^n \frac{t^{(2n+1)\alpha}}{\Gamma(1 + (2n+1)\alpha)} \tag{6.25} $$

$$ {}_{\mathrm{C}}\cos(\alpha, t) = \sum_{n=0}^{\infty} (-1)^n \frac{t^{2n\alpha}}{\Gamma(1 + 2n\alpha)} \tag{6.26} $$

The major property of these series is:

$$ {}_{\mathrm{C}}D^{\alpha}\, {}_{\mathrm{C}}\sin(\alpha, \omega t) = \omega^{\alpha}\, {}_{\mathrm{C}}\cos(\alpha, \omega t) \tag{6.27} $$

$$ {}_{\mathrm{C}}D^{\alpha}\, {}_{\mathrm{C}}\cos(\alpha, \omega t) = -\omega^{\alpha}\, {}_{\mathrm{C}}\sin(\alpha, \omega t) \tag{6.28} $$

They are directly related to the Mittag-Leffler function (2.17):

$$ {}_{\mathrm{C}}\sin(\alpha, t) = t^{\alpha} E(2\alpha, 1 + \alpha, -t^{2\alpha}) \tag{6.29} $$

$$ {}_{\mathrm{C}}\cos(\alpha, t) = E(2\alpha, -t^{2\alpha}) \tag{6.30} $$

In Fig. 6.3 we have sketched a graphical representation.

Let us compare the derived solutions of the harmonic oscillator differential equation for different definitions of the fractional derivative. First we observe a surprising similarity of the presented solutions. All of them decay exponentially for $\alpha < 1$ which may be interpreted from a classical point of view as a damping phenomenon.

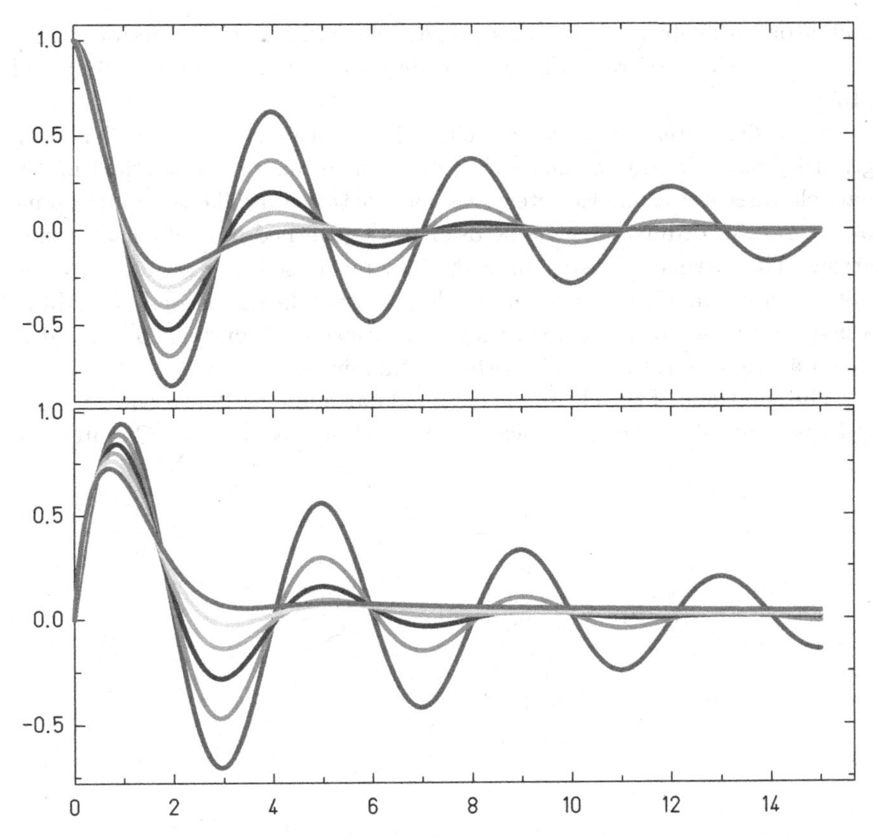


Fig. 6.3. Solutions $_C\cos(\alpha, t)$ and $_C\sin(\alpha, t)$ of the harmonic oscillator differential equation with Caputo derivative with $\alpha = 0.95, 0.90, 0.85, 0.80, 0.75, 0.70$.

Furthermore the frequency of these damped oscillations for a given $\alpha \approx 1$ are quite similar for the different fractional derivative definitions presented and therefore may be regarded as a definition-independent property.

The main difference for the different solutions is the number of zeroes. This will be an important observation for solutions of fractional wave-equations when appropriate boundary conditions are imposed.

In general the fractional harmonic oscillator behaves like a damped harmonic oscillator for $\alpha < 1$. The damping is intrinsic to the fractional oscillator and not due to external influences such as friction [Mainardi(1996), Gorenflo(1996), Tofighi(2003), Stanislavsky(2006a,b), Yonggang(2010)]. Nevertheless, there is a correspondence of the solutions of the fractional oscillator with solutions of a classical particle with fractional friction, which becomes obvious, if we compare the determining polynomials (6.8) and (4.22).

A physical interpretation for the fractional harmonic oscillator has been given by Stanislavsky [Stanislavsky(2004)] in terms of a statistical multiparticle interpretation: He interprets the fractional oscillator as an ensemble average of ordinary harmonic oscillators governed by a stochastic time arrow. The intrinsic absorption of the fractional oscillator results from the full contribution of the harmonic oscillator ensemble: these oscillators differ a little from each other in frequency so that each response is compensated by an anti-phase response of another harmonic oscillator.

The solutions of the classical fractional harmonic oscillator will turn out to be helpful solving the free fractional Schrödinger equation in Chapter 8.3.

# Chapter 7

# Wave Equations and Parity

*The classical field of application for wave equations is the determination of the eigenfrequencies and eigenfunctions of vibrating systems.*

*Obviously there is a close analogy with the differential equation of the harmonic oscillator.*

*Therefore an extension of wave equations to the fractional case will benefit from solutions presented in the previous chapter.*

*A new aspect is the use of boundary conditions in contrast to initial conditions. Furthermore, since a wave equation besides the time dependence also contains a spatial part, we are forced to consider the behaviour of solutions for negative arguments.*

*For free solutions, this problem may be solved by the additional requirement, that these solutions should be eigenfunctions of the parity operator, too.*

## 7.1 Fractional wave equations

A possible extension of a classical wave equation (6.3) to the fractional case is given by

$$\left(\frac{d^{2\alpha}}{dx^{2\alpha}} + \frac{d^{2\alpha}}{dy^{2\alpha}} + \frac{d^{2\alpha}}{dz^{2\alpha}} - \mu^2 \frac{d^2}{dt^2}\right) \psi(x, y, z, t) = 0 \tag{7.1}$$

Here we have extended only the spatial contribution to the fractional case, while the time derivative remains unchanged being of second order. The dimension of the parameter $\mu$ is determined in the mks-system by $[s/m^{\alpha}]$.

Because (3.14) holds, a separation of variables is possible using the product ansatz:

$$\psi(x, y, z, t) = X(x)Y(y)Z(z)T(t) \tag{7.2}$$

The partial differential equation (7.1) is therefore transformed into a system of uncoupled ordinary differential equations:

$$\frac{d^{2\alpha}}{dx^{2\alpha}}X = -\kappa_x^2 X \tag{7.3}$$

$$\frac{d^{2\alpha}}{dx^{2\alpha}}Y = -\kappa_y^2 Y \tag{7.4}$$

$$\frac{d^{2\alpha}}{dx^{2\alpha}}Z = -\kappa_y^2 Z \tag{7.5}$$

$$\frac{d^2}{dx^2}T = \frac{\omega^2}{\mu^2}T \tag{7.6}$$

The separation constants $\omega, \kappa_x, \kappa_y, \kappa_z$ fulfil the condition

$$\omega = \mu\sqrt{\kappa_x^2 + \kappa_y^2 + \kappa_z^2} \tag{7.7}$$

A solution of this system of fractional differential equations may be determined, if an appropriate set of boundary conditions (Dirichlet, Neumann, mixed) at the boundaries of a given domain is known.

The simple case of a vibrating string with length $L = 2a$, which is fixed at its ends for example is described classically by a one dimensional wave equation for the oscillation $X(x)$ and is uniquely determined by the Dirichlet boundary conditions:

$$X(-a) = X(a) = 0 \tag{7.8}$$

The differential equation for a vibrating string is equivalent to the differential equation for the harmonic oscillator, where from the homogeneous set of all possible solutions those are selected, which obey the boundary conditions (7.8).

This strategy may also be directly applied in the case of a fractional wave equation. But this implies, that we are forced to make statements on the behaviour of solutions for negative argument too. Until now, for the fractional harmonic oscillator we have only presented solutions as a function of the variable $x$ for $x > 0$.

We will solve this problem by introducing additional assumptions on the behaviour of solutions under space inversion.

## 7.2 Parity and time-reversal

The transition of variables and derivatives from integer to fractional values

$$\left\{x^n, \frac{d^n}{dx^n}\right\} \to \left\{x^\alpha, \frac{d^n}{dx^\alpha}\right\} \tag{7.9}$$

makes it necessary to consider the behaviour of functions under space inversions.

We present a simple example to clarify the problem: The function

$$f(x) = x \tag{7.10}$$

behaves under space inversions as

$$f(-x) = -x = -f(x) \tag{7.11}$$

and therefore has negative parity or in other words is an odd function. If we choose the simple substitution $x \to x^\alpha$, we obtain for the fractional function

$$f(x) = x^\alpha \tag{7.12}$$

$$f(-x) = (-x)^\alpha = (-1)^\alpha |x|^\alpha \tag{7.13}$$

a multivalent value (e.g. for the semiderivative $\alpha = 1/2$ we have two different purely imaginary function values $\pm i|x|^{\frac{1}{2}}$). The parity property is lost.

If we intend to conserve the parity properties of normal functions for the corresponding fractional generalization, we must specify the parity behaviour for the variable $x$.

With the ansatz

$$x \to \text{sign}(x)|x|^\alpha = \hat{x}^\alpha \tag{7.14}$$

$$\frac{d}{dx} \to \frac{d^\alpha}{d\hat{x}^\alpha} = \hat{D}_x^\alpha \tag{7.15}$$

this problem is solved.

Instead of (7.1) we will therefore use

$$\left(\left(\frac{d^\alpha}{d\hat{x}^\alpha}\right)^2 + \left(\frac{d^\alpha}{d\hat{y}^\alpha}\right)^2 + \left(\frac{d^\alpha}{d\hat{z}^\alpha}\right)^2 - \mu^2 \frac{d^2}{dt^2}\right) \psi(x, y, z, t) = 0 \tag{7.16}$$

as a regularized fractional wave equation. The free solutions of this wave equation show then a similar behaviour under parity transformations for arbitrary $\alpha$ as in the limiting case $\alpha = 1$.

This regularization procedure is equivalent to the use of the left and right handed fractional derivative definitions according to Riemann and Caputo, because these are distinguished at $x = 0$. For Liouville type derivative definitions the regularization is an additional requirement, since from the original definition for left- and right-handed Liouville type derivatives the point $x = 0$ is not distinguished.

Until now we have only discussed the spatial part of a fractional equation. If we allow a fractional derivative for the time variable too, a similar problem arises for time-reversal.

With the ansatz

$$t \to \mathrm{sign}(t)|t|^\alpha = \hat{t}^\alpha \tag{7.17}$$

$$\frac{d}{dt} \to \frac{d^\alpha}{d\hat{t}^\alpha} = \hat{D}_t^\alpha \tag{7.18}$$

the function $g(t) = (\hat{t}^\alpha)^{2n}$ behaves as

$$g(-t) = (\mathrm{sign}(-t)|-t|^\alpha)^{2n} = (-1)^{2n}(\mathrm{sign}(t)|t|^\alpha)^{2n} = g(t) \tag{7.19}$$

and the function $u(t) = (\hat{t}^\alpha)^{2n+1}$ behaves as

$$u(-t) = (\mathrm{sign}(-t)|-t|^\alpha)^{2n+1} = (-1)^{2n+1}(\mathrm{sign}(t)|t|^\alpha)^{2n+1} = -u(t) \tag{7.20}$$

and consequently g(t) and u(t) are eigenfunctions of the time-reversal operator. Free solutions of such a regularized time fractional derivative equation show a similar behaviour under time-reversal as for $\alpha = 1$.

A simultaneous conservation of parity and time-reversal may be achieved, if space and time coordinates are treated similarly. Of course, these are symmetries, which are intentionally imposed. Interrelations between the non conservation properties of parity, time reversal, and charge conjugation are discussed e.g. in [Lee(1957)].

## 7.3 Solutions of the free regularized fractional wave equation

The solutions of the free regularized one dimensional fractional wave equation

$$\left( \left( \frac{d^\alpha}{d\hat{x}^\alpha} \right)^2 + \kappa_x^2 \right) X(x) = 0 \tag{7.21}$$

are given in the sense of (7.14) using the regularized fractional derivative according to Fourier as the solutions for the fractional harmonic oscillator (6.14) and (6.15) extended to negative arguments:

$$_{\mathrm{F}}\sin(\alpha, x) = \frac{1}{2i}\left(e^{\frac{1}{2}e^{\frac{i\pi}{2\alpha}}|x|} - e^{\frac{1}{2}e^{-\frac{i\pi}{2\alpha}}|x|}\right) \tag{7.22}$$

$$_{\mathrm{F}}\cos(\alpha, x) = \mathrm{sign}(x)\frac{1}{2}\left(e^{\frac{1}{2}e^{\frac{i\pi}{2\alpha}}|x|} + e^{\frac{1}{2}e^{-\frac{i\pi}{2\alpha}}|x|}\right) \tag{7.23}$$

For the fractional derivative according to Riemann solutions are given as a spatial extension of (6.21) and (6.22):

$$_{\mathrm{R}}\sin(\alpha, x) = \mathrm{sign}(x)|x|^{2\alpha-1}E(2\alpha, 2\alpha, -|x|^{2\alpha}) \tag{7.24}$$

$$_{\mathrm{R}}\cos(\alpha, x) = |x|^{\alpha-1}E(2\alpha, \alpha, -|x|^{2\alpha}) \tag{7.25}$$

For the fractional Caputo derivative the regularized solutions are given as a spatial extension of (6.29) and (6.30):

$$_{\mathrm{C}}\sin(\alpha, x) = \mathrm{sign}(x)|x|^{\alpha}E(2\alpha, 1+\alpha, -|x|^{2\alpha}) \tag{7.26}$$

$$_{\mathrm{C}}\cos(\alpha, x) = E(2\alpha, -|x|^{2\alpha}) \tag{7.27}$$

Application of the regularized fractional derivative leads to:

$$\hat{D}^{\alpha}\ _{\mathrm{F,R,C}}\sin(\alpha, \kappa_x x) = \mathrm{sign}(\kappa_x)|\kappa_x|^{\alpha}\ _{\mathrm{F,R,C}}\cos(\alpha, \kappa_x x) \tag{7.28}$$

$$\hat{D}^{\alpha}\ _{\mathrm{F,R,C}}\cos(\alpha, \kappa_x x) = -\mathrm{sign}(\kappa_x)|\kappa_x|^{\alpha}\ _{\mathrm{F,R,C}}\sin(\alpha, \kappa_x x) \tag{7.29}$$

Obviously the function $\sin(\alpha, x)$ is odd while $\cos(\alpha, x)$ is an even function:

$$_{\mathrm{F,R,C}}\sin(\alpha, -x) = -\,_{\mathrm{F,R,C}}\sin(\alpha, x) \tag{7.30}$$

$$_{\mathrm{F,R,C}}\cos(\alpha, -x) = \,_{\mathrm{F,R,C}}\cos(\alpha, x) \tag{7.31}$$

Therefore parity is conserved even for $\alpha \neq 1$.

For a given set of boundary conditions of type (7.8) it is of fundamental importance to determine the zeroes of the proposed solutions:

The zeroes for the solutions of the one dimensional wave equation with fractional derivative according to Fourier are given by:

$$_{\mathrm{F}}x_0(n, \alpha) = \frac{(n+1)}{\sin(\frac{\pi}{2a})}\frac{\pi}{2} \qquad n = 0, 1, 2, 3... \tag{7.32}$$

The zeroes for solutions based on the Riemann and Caputo derivative can only be determined numerically and are listed in Table 7.1.

In Fig. 7.1 we have plotted the zeroes of the functions $_{\mathrm{F,R,C}}\sin(\alpha, x)$ and $_{\mathrm{F,R,C}}\cos(\alpha, x)$. While position differences of the zeroes for $\alpha > 1$ are minimal, they are steadily increasing for $\alpha < 1$.

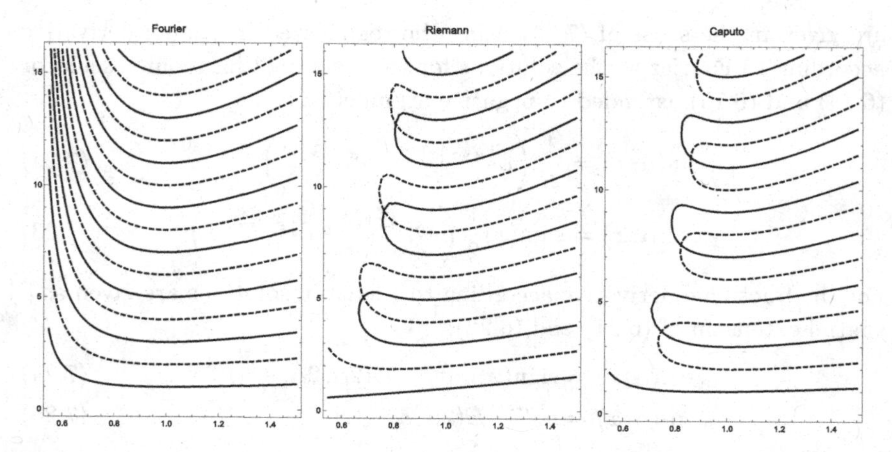


Fig. 7.1. Zeroes of solutions of the fractional one dimensional wave equation with good parity for $0.55 \leq \alpha \leq 1.5$ in units $\pi/2$. Thick lines indicate the zeroes of $\cos(\alpha, \pi x/2)$. Dashed lines indicate the zeroes of $\sin(\alpha, \pi x/2)$. From left to right results are given for the Fourier-, Riemann- and Caputo fractional derivative.

The solutions for the Fourier derivative in the case $\alpha < 1$ are equidistant and countable infinite, for the solutions based on the Riemann and Caputo derivative definition there exist only a finite number of zeroes. For $\alpha < 0.65$ using the Riemann derivative there exists only one zero for the lowest even and odd solution. For the Caputo derivative in this region there exists only a single zero for the lowest even solution.

If we apply our results to the example of a vibrating string, for solutions of the fractional wave equation using the Fourier definition of a fractional derivative an arbitrary number of overtones is possible and the corresponding spectrum differs only by a multiplicative constant from the case $\alpha = 1$. Using the Riemann- and Caputo derivative for $\alpha < 1$ there is only a limited number of overtones.

For $\alpha > 1$ there is no limitation for the number of zeroes any more. In Fig. 7.3 the solutions for $\alpha = 1.1$ are presented. We observe, that the graphs do not differ too much for a specific choice of a fractional derivative.

A comparison of the solutions with the standard trigonometric functions sine and cosine respectively leads to the result, that for $\alpha > 1$ the solutions are concentrated at the boundaries of a given domain, while for $\alpha < 1$ the solutions are centred within a given domain.

Since wave equations play a dominant role in many different branches of physics, the results derived for the classical harmonic oscillator are of

Table 7.1. The first eight zeroes for the solutions of the fractional one dimensional wave equation for Fourier-, Riemann- and Caputo fractional derivative in units $\pi/2$ and given parity.

| $\alpha$ | $n^P$ | ${}_{\mathrm{F}}D^\alpha$ | ${}_{\mathrm{R}}D^\alpha$ | ${}_{\mathrm{C}}D^\alpha$ |
|---|---|---|---|---|
| 0.95 | $0^+$ | 1.00343 | 0.94855 | 0.99386 |
| | $1^-$ | 2.00685 | 1.95190 | 2.01681 |
| | $2^+$ | 3.01028 | 2.95878 | 3.01353 |
| | $3^-$ | 4.01371 | 3.96151 | 4.00558 |
| | $4^+$ | 5.01714 | 4.96351 | 5.01525 |
| | $5^-$ | 6.02056 | 5.96746 | 6.02815 |
| | $6^+$ | 7.02399 | 6.97178 | 7.02536 |
| | $7^-$ | 8.02742 | 7.97478 | 8.01976 |
| 0.90 | $0^+$ | 1.01543 | 0.89801 | 0.99286 |
| | $1^-$ | 2.03085 | 1.91216 | 2.05892 |
| | $2^+$ | 3.04628 | 2.93753 | 3.05684 |
| | $3^-$ | 4.06171 | 3.95115 | 4.03050 |
| | $4^+$ | 5.07713 | 4.96090 | 5.06866 |
| | $5^-$ | 6.09256 | 5.97828 | 6.13263 |
| | $6^+$ | 7.10799 | 6.99857 | 7.11648 |
| | $7^-$ | 8.12341 | 8.01178 | 8.06917 |
| 0.85 | $0^+$ | 1.03969 | 0.84848 | 0.99901 |
| | $1^-$ | 2.07938 | 1.88358 | 2.14271 |
| | $2^+$ | 3.11907 | 2.94561 | 3.14622 |
| | $3^-$ | 4.15876 | 3.98181 | 4.06146 |
| | $4^+$ | 5.19845 | 5.00326 | 5.16727 |
| | $5^-$ | 6.23814 | 6.04877 | 6.43227 |
| | $6^+$ | 7.27783 | 7.11094 | 7.32333 |
| | $7^-$ | 8.31752 | 8.14094 | 7.98882 |

general interest and give a first insight into the structure of solutions of fractional wave equations.

Until now, we have presented solutions which fulfil boundary conditions (7.8) which are synonymous to a determination of zeroes. In the next section we will demonstrate, that in the case of a fractional differential equation a given set of boundary conditions determines the behaviour of the solution not only within a given domain, but on $-\infty < x < +\infty$ in general.

Table 7.1. (*Continued*)

| $\alpha$ | $n^P$ | ${}_{\mathrm{F}}D^{\alpha}$ | ${}_{\mathrm{R}}D^{\alpha}$ | ${}_{\mathrm{C}}D^{\alpha}$ |
|---|---|---|---|---|
| 0.80 | $0^+$ | 1.08239 | 0.80009 | 1.01544 |
| | $1^-$ | 2.16478 | 1.87053 | 2.30691 |
| | $2^+$ | 3.24718 | 3.00027 | 3.31519 |
| | $3^-$ | 4.32957 | 4.07696 | 4.03669 |
| | $4^+$ | 5.41196 | 5.10097 | 5.29765 |
| | $5^-$ | 6.49435 | 6.19970 | - |
| | $6^+$ | 7.57675 | 7.39569 | 7.88035 |
| | $7^-$ | 8.65914 | 8.43172 | - |
| 0.75 | $0^+$ | 1.1547 | 0.75299 | 1.04739 |
| | $1^-$ | 2.3094 | 1.88016 | 2.70995 |
| | $2^+$ | 3.4641 | 3.13874 | 3.65656 |
| | $3^-$ | 4.6188 | 4.28656 | 3.71284 |
| | $4^+$ | 5.7735 | 5.23632 | 5.33263 |
| | $5^-$ | 6.9282 | 6.43402 | - |
| | $6^+$ | 8.0829 | - | - |
| | $7^-$ | 9.2376 | 9.29997 | - |
| 0.70 | $0^+$ | 1.27905 | 0.70734 | 1.10430 |
| | $1^-$ | 2.55810 | 1.92522 | - |
| | $2^+$ | 3.83714 | 3.47433 | - |
| | $3^-$ | 5.11619 | 4.76139 | - |
| | $4^+$ | 6.39524 | 5.22151 | - |
| | $5^-$ | 7.67429 | - | - |
| | $6^+$ | 8.95334 | - | - |
| | $7^-$ | 10.2324 | - | - |
| 0.67 | $0^+$ | 1.39792 | 0.68073 | 1.15767 |
| | $1^-$ | 2.79585 | 1.97882 | - |
| | $2^+$ | 4.19377 | 4.06990 | - |
| | $3^-$ | 5.59169 | - | - |
| | $4^+$ | 6.98962 | 4.73238 | - |

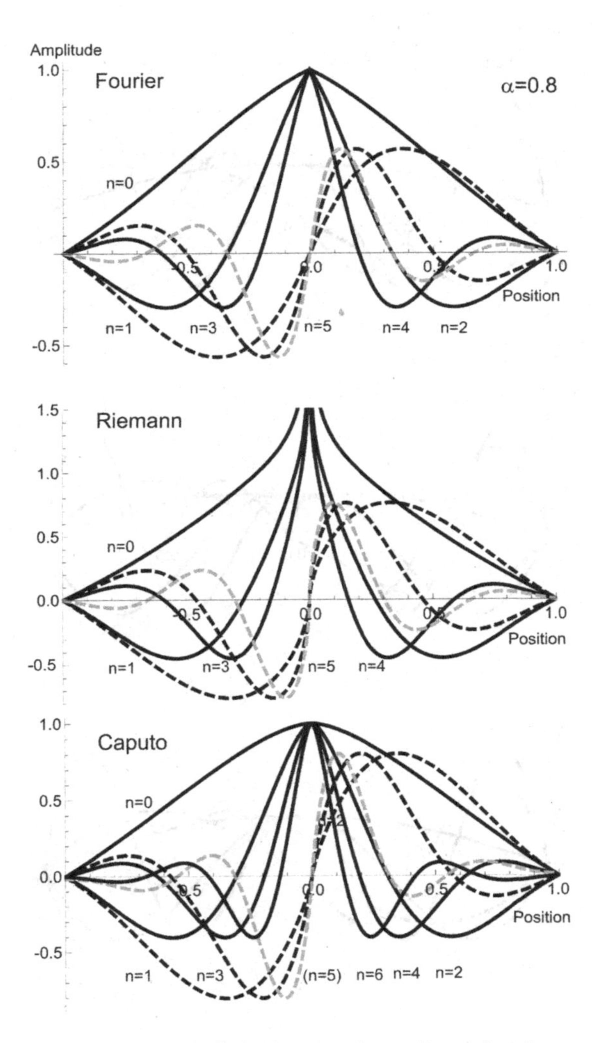


Fig. 7.2. A plot of the solutions of the fractional one dimensional wave equation with good parity for $\alpha = 0.8$. Even parity solutions using the Riemann derivative are singular at $x = 0$ for $\alpha < 1$.

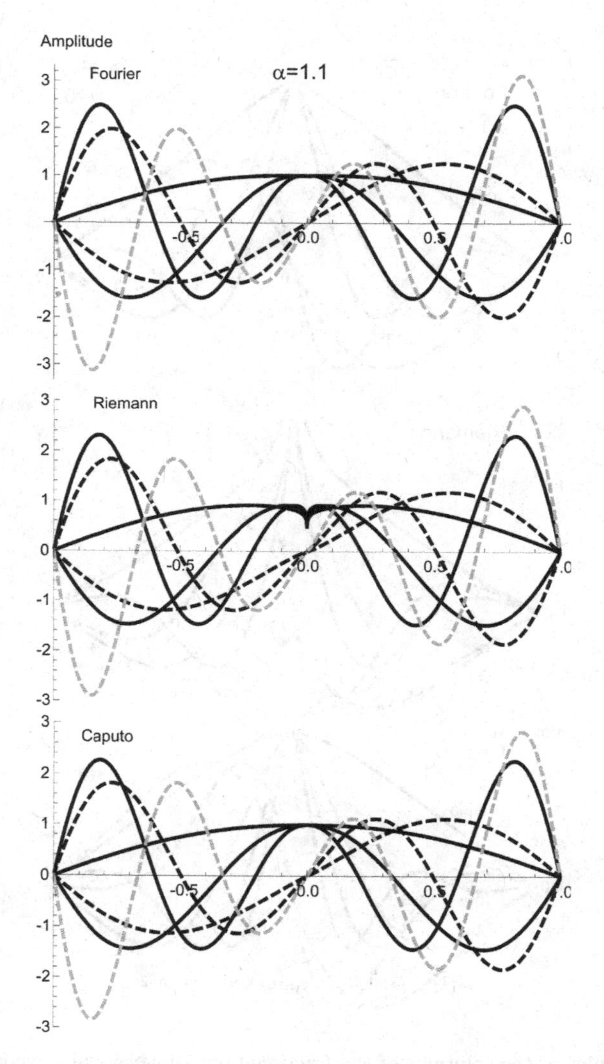


Fig. 7.3. A plot of the solutions of the fractional one dimensional wave equation with good parity for $\alpha = 1.1$. Solutions with even parity using the Riemann derivative for $\alpha > 1$ are passing the origin.

## Chapter 8

# Nonlocality and Memory Effects

*Now we will embed the concept of fractional calculus into a broader theory of nonlocal operators. This strategy will lead to a successful generalization of the fractional derivative operator and will result in a promising concept for a deduction of a covariant fractional derivative calculus on the Riemann space in the future.*

*We will demonstrate the influence of a memory effect for a distinguished set of histories presenting a full analytic solution for a simple nonlocal model for a metal hydride battery.*

## 8.1 A short history of nonlocal concepts

The concept of nonlocality and memory effects has a long historical tradition in physics.

Figure 8.1 illustrates two examples for nonlocal concepts. On the left a woodcut from the 1551 edition of *Rudimenta mathematica* [Münster(1551)] is reproduced showing the ballistic curve of a canon ball, which at that time was determined on the basis of the impetus theory, which was the fundamental pre-Newtonian dynamic theory of motion for more than 1000 years.

According to Leonardo da Vinci impetus "is the impression of local movement transmitted from the mover to the mobile object and maintained by the air or by the water as they move to prevent a vacuum" [da Vinci(1488)]. A logical consequence of this definition is the prediction of a curved movement of a mobile object, if this object moves in a circle and is released in the course of this motion, which is known as circular impetus. In terms of a modern nonlocal theory such a behaviour was a direct consequence of a memory effect: The actual state of the object contains a

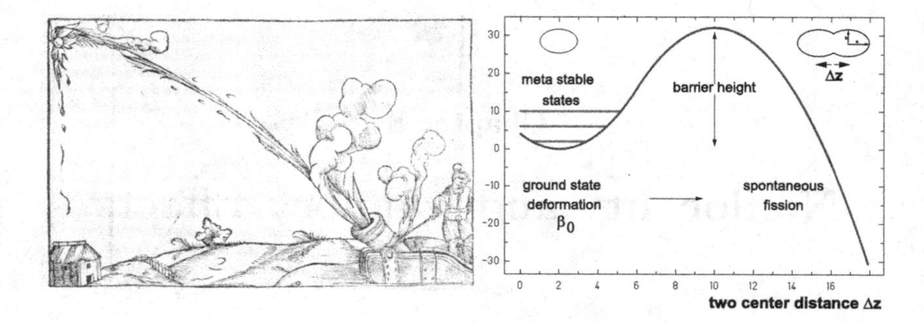


Fig. 8.1. Two illustrative examples for nonlocal concepts in physics.

circular component since it is influenced by an earlier circular state in the past. Following Newton, this conclusion is wrong and in course of Newton's theory's continued success the medieval impetus theory was dead.

But at the beginning of the twentieth century atomic and sub atomic physics caused an unexpected triumphant return of nonlocal concepts in terms of quantum mechanics.

A prominent example is the tunnel effect, which cannot be explained within the framework of classical physics. A typical application is spontaneous fission, where a nucleus disintegrates into two or more large fragments. The fission process in general was first proposed by Ida Noddack [Noddack(1934)] as an alternative interpretation of Fermi's experimental results [Fermi(1934)]. The spontaneous fission of uranium was first observed by Flerov [Flerov(1940)].

In Fig. 8.1 on the right this process is illustrated schematically. As a function of the total elongation of the nucleus ($\Delta z$) a collective potential is shown. The state of the nucleus is determined solving the collective Schrödinger equation including this potential. The meta stable ground state of the nucleus is located at a given ground state deformation $\beta_0$ which determines a local minimum in the potential function. From a classical point of view, the nucleus could stay there for ever.

But as a consequence of Heisenberg's uncertainty relation there is a non vanishing probability for a separated configuration beyond the finite potential barrier. As a result there is a finite lifetime $T$ for the compound nucleus which is given in WKB-approximation [Martin(2006)] as

$$\log(T) \approx \frac{2}{\hbar} \int d(\Delta z) \sqrt{2mE - V(\Delta z)}. \tag{8.1}$$

In view of a nonlocal theory, the time development of the meta-stable ground state is influenced by the fission potential characteristics for large $\Delta z$ values, which are not accessible from a classical point of view.

Consequently, the figures presented in Fig. 8.1 mark a remarkable historical development of nonlocal concepts in physics. And finally, modern string theory could be reckoned as one more nonlocal extension of a point particle dynamics. Within that context, fractional calculus is a revival, sublimation and continuation of nonlocal concepts developed in the past.

Until now we have proceeded with the presentation of the conceptual basis of fractional calculus along a historical path, which started with the realization of the fractional derivative for special function classes followed by an extension to general analytic function sets based on Cauchy's integral formula.

Now we intend to embed the fractional calculus into a much broader context of a theory of nonlocal extensions of standard operators.

We will first present a gradual classification scheme of different levels of nonlocality. We will then investigate some of the general properties of a certain class of normalizable nonlocal operators and we will determine their relation to fractional operators. Finally we will apply the concept of nonlocality to demonstrate the influence of memory effects on the solution of nonlocal differential equations.

## 8.2 From local to nonlocal operators

In the following we present a concept of nonlocality, which is based on the increasing amount of information which has to be gathered in the neighbourhood of a given local quantity to obtain a specific realization of the corresponding nonlocal quantity.

The position of a point mass is given in terms of classical mechanics according to Newton for a given time t as $x(t)$. It is given in the mks-system in units of $[m]$. This implies that there exists a reference position, e.g. a zero-point of a space coordinate system. The position of a point mass is measured with respect to this reference position. Since the position for a given time t is independent from a position at $t+\delta t$ and $t-\delta t$ respectively, we call $x(t)$ a purely local quantity or in other words a zero-nonlocal quantity, because information within a circle of radius $r = 0$ centred at $x(t)$ suffices to determine the position (neglecting the work of the geometer who devoted his life to establish a precisely gauged coordinate system). Besides the position are many more variables of this locality type, e.g. temperature, pressure, field strength et cetera.

The velocity of a point mass is given according to Newton by the first derivative of the position $x(t)$ with respect to $t$.

Following Weierstraß the first derivative may be defined as

$$v(t) = \frac{d}{dt}x(t) = \begin{cases} \lim_{h\to 0} \frac{x(t)-x(t-h)}{h} & \text{left, causal} \\ \lim_{h\to 0} \frac{x(t+h)-x(t-h)}{2h} & \text{symmetric} \\ \lim_{h\to 0} \frac{x(t+h)-x(t)}{h} & \text{right, anti-causal,} \quad h < \epsilon, h \geq 0 \end{cases} \tag{8.2}$$

where $\epsilon$ is an arbitrary small positive real value.

As long as $x(t)$ is an analytic smooth function of $t$ the three given definitions seem to be equivalent. As long as $t$ is interpreted as a space-like coordinate, the definitions indicate that position information is gathered from the left only, from the right only or from both sides simultaneously.

If $t$ is interpreted as a time-like coordinate, the first definition of (8.2) obeys the causality principle, which roughly states, that the current state of a physical object may only be influenced by events, which already happened in the past (concepts of past, future, causality and locality are discussed e.g. in [Kant(1781), Einstein(1935), Weizsäcker(1939), Treder(1983), Todd(1992), Rowling(2000)]).

The last of definitions (8.2) clearly violates the principle of causality within the framework of a classical context. Of course, in the area of quantum mechanics we may adopt Stückelberg's [Stückelberg(1941)] and Feynman's [Feynman(1949)] view and recall their idea, that anti-particles propagate backwards in time. Therefore this definition is appropriate to describe e.g. the velocity of an anti-particle.

Finally the second definition in (8.2) of a derivative is a mixture of causal and anti-causal propagation, which indicates, that a single particle interpretation may lead to difficulties. Therefore it should be considered, if this definition could be used to describe the velocity of a particle anti-particle pair.

Though for ordinary derivatives this discussion sounds somewhat sophisticated and artificial, since whatever definition we use, a violation of the causality principle is only of order $\epsilon$, in case of really nonlocal quantities these considerations become important.

Since we need information within a radius of $r < \epsilon$ centred at $x(t)$ in order to determine the standard derivative we call such a quantity to be $\epsilon$-nonlocal. In standard textbooks the ordinary derivative is classified as a local quantity, since $\epsilon$ is arbitrarily close to zero. Therefore we may use the terms $\epsilon$-nonlocal and local simultaneously, as long as $\epsilon \approx 0$.

Besides velocity there are many more variables of this locality type, e.g. currents, flux densities and last not least all variables, which are determined by a differential equation. As a next step we will reformulate the ordinary derivative in terms of an integral form

$$v(t) = \frac{d}{dt}x(t) = \begin{cases} 2\int_0^\infty dh\,\delta(h)\,v(t-h) & \text{left, causal} \\ 2\int_0^\infty dh\,\delta(h)\,\frac{v(t+h)+v(t-h)}{2} & \text{symmetric} \\ 2\int_0^\infty dh\,\delta(h)\,v(t+h) & \text{right, anti-causal} \end{cases} \tag{8.3}$$

where $\delta(t)$ is the Dirac-delta function which has the property [Lighthill(1958)]:

$$\int_{-\infty}^{\infty} dh\,\delta(h) f(h) = f(0) \tag{8.4}$$

and is defined as a class of all equivalent regular sequences of good functions $w(a,t)$

$$\delta(t) = \lim_{a\to 0} w(a,t) \tag{8.5}$$

where $a$ is a smooth parameter with $a \geq 0$.

Following [Lighthill(1958)] a good function is one, which is everywhere differentiable any number of times and such, that it and all its derivatives are $O(|t|^{-N})$ as $|t| \to \infty$ for all $N$. Furthermore a sequence $w(a,t)$ of good functions is called regular, if, for any good function $f(t)$ the limit

$$\lim_{a\to 0} \int_{-\infty}^{+\infty} w(a,t) f(t) dt \tag{8.6}$$

exists. Finally two regular sequences of good functions are called equivalent if, for any good function $f(t)$ whatever, the limit (8.6) is the same for each sequence.

Typical examples for such sequences of good functions are

$$w(a,t) = \begin{cases} \exp^{-(|t|/a)^p} & \text{exponential, } p \in \mathbb{R}_+ \\ \mathrm{Ai}(|t|/a) & \text{Airy} \\ \frac{1}{t}\sin(|t|/a) & \text{sine} \end{cases} \tag{8.7}$$

In Fig. 8.2 we have collected some graphs of good functions.

If the smooth parameter $a$ is close to zero, the relations

$$\lim_{a\to 0} 2\int_0^\infty dh\, w(a,h) v(t\pm h) \approx \lim_{a\to 0} 2\int_0^\epsilon dh\, w(a,h) v(t\pm h) = v(t) \tag{8.8}$$

become valid and therefore definition (8.3) is equivalent to (8.2) for $a < \epsilon$. Consequently for this case we need information within a radius $r < \epsilon$ centred at $x(t)$ and the definition (8.3) leads to a $\epsilon$-nonlocality.

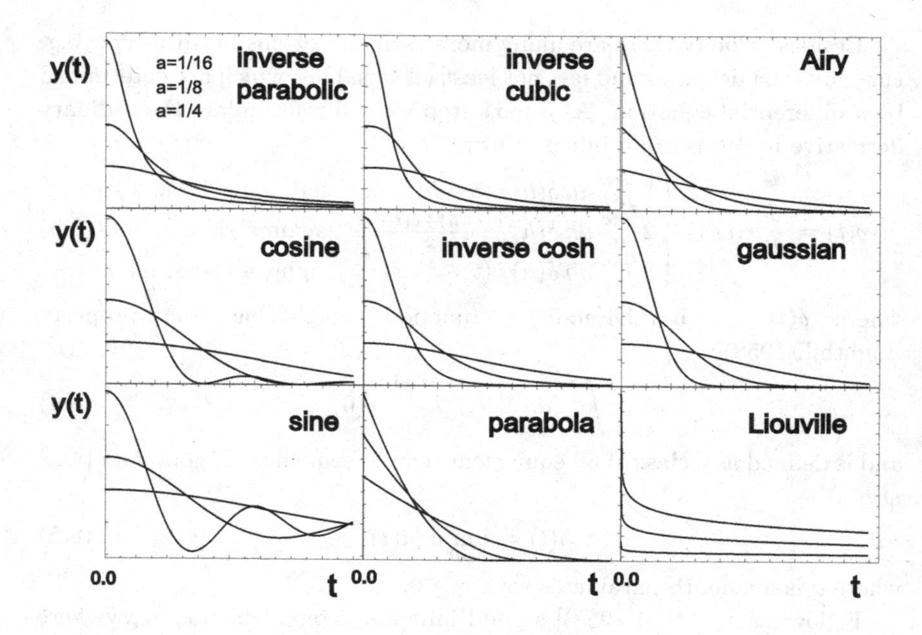


Fig. 8.2. Weight functions $w(a,t)$ listed in Table 8.1 for different values of $a$.

But if we allow that $a$ may be any finite real value, we have to gather information within a radius of arbitrary size, therefore

$$\frac{d}{dt}_{\text{nonlocal}} x(t) = \begin{cases} \frac{1}{N}\int_0^\infty dh\, w(a,h)\, v(t-h) & \text{left, causal} \\ \frac{1}{N}\int_0^\infty dh\, w(a,h)\, \frac{v(t+h)+v(t-h)}{2} & \text{symmetric} \\ \frac{1}{N}\int_0^\infty dh\, w(a,h)\, v(t+h) & \text{right, anti-causal} \end{cases} \tag{8.9}$$

with

$$\lim_{a\to\infty} w(a,t) = \delta(t) \tag{8.10}$$

and a normalization constant $N$ which is determined by the condition

$$\frac{1}{N}\int_0^\infty dh\, w(a,h) = 1 \tag{8.11}$$

is a reasonable extension from the local operator $\frac{d}{dt}$ to a nonlocal version of the same operator.

Let us introduce a set of shift operators $\hat{s}(h)$ which acts on an analytic

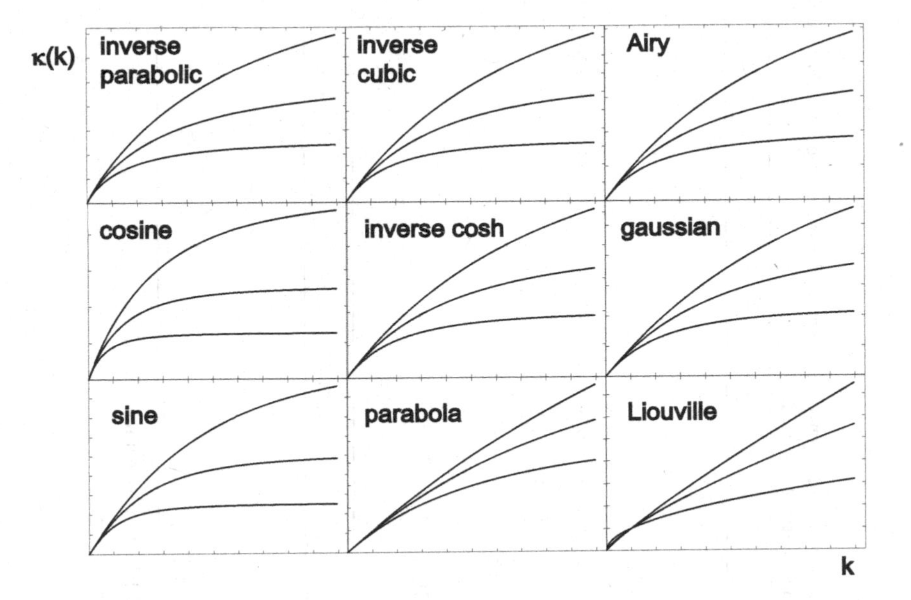


Fig. 8.3. Eigenvalue spectrum (8.48) of the nonlocal first order differential equation (8.43) for different weight functions $w(a,t)$ for different values of $a$. Function names are listed in Table 8.1.

function $f(x)$ like:

$$\hat{s}_+ f(x) = f(x+h) \tag{8.12}$$

$$\hat{s}_- f(x) = f(x-h) \tag{8.13}$$

$$\frac{1}{2}(\hat{s}_+ + \hat{s}_-) f(x) = \frac{1}{2}\left(f(x+h) + f(x-h)\right) \tag{8.14}$$

$$\frac{1}{2h}(\hat{s}_+ - \hat{s}_-) f(x) = \frac{1}{2h}\left(f(x+h) - f(x-h)\right) = \frac{\Delta f}{\Delta x} \tag{8.15}$$

where the last operation is equivalent to the secant approximation of the first derivative of $f(x)$.

A possible generalization of the proposed procedure for an arbitrary local operator $\hat{O}_{\text{local}}(x)$ to the corresponding nonlocal representation $\hat{O}_{\text{nonlocal}}(x)$ is then given according to (8.9) by:

$$\hat{O}_{\text{nonlocal}}(x)\, f(x) = \frac{1}{N}\int_0^\infty dh\, w(a,h)\, \hat{s}(h)\, \hat{O}_{\text{local}}(x)\, f(x) \tag{8.16}$$

Therefore the transition from a local representation of a given operator to a reasonable nonlocal representation of the same operator is given by three operations:

Table 8.1. A listing of possible weight functions $w(a,h)$ with name, definition, norm, and eigenvalue spectrum for the first order differential equation $\frac{d^a}{dx^a}y(x) = \kappa(k)y(x)$. In the last column function names according to Chapter 2 are used.

| name | $w(a,h)$ | $N$ | $\kappa(k)$ |
|---|---|---|---|
| inverse parabolic | $\frac{1}{1+(h/a)^2}$ | $\frac{a\pi}{2}$ | $\frac{k}{\pi}(2\sin(ak)\mathrm{Ci}(ak) + \cos(ak)(\pi - 2\mathrm{Si}(ak)))$ |
| inverse cubic | $\frac{1}{1+(h/a)^3}$ | $\frac{2\pi}{3\sqrt[3]{a}}$ | $\frac{3k}{4\pi^2} G^{4,1}_{1,4}\left(\frac{a^3k^3}{27}\middle\vert \begin{matrix}\frac{2}{3}\\ 0, \frac{1}{3}, \frac{2}{3}, \frac{2}{3}\end{matrix}\right)$ |
| Airy | $\mathrm{Ai}(x/a)$ | $\frac{a}{3}$ | $-\frac{ak^2}{2 3^{5/6}\pi} e^{-a^3k^3/3}(\mathrm{Ei}_{2/3}(-(ka)^3/3)\Gamma(-1/3) + 3^{2/3}ak\mathrm{Ei}_{1/3}(-(ka)^3/3)\Gamma(1/3))$ |
| cosine | $\frac{1}{h^2}(1-\cos(h/a))$ | $\frac{\pi}{2a}$ | $\frac{k}{\pi}(2\mathrm{ArcCot}(ak) - ak\log(1+\frac{1}{ka}))$ |
| inverse cosh | $\cosh(h/a)^{-2}$ | $a$ | $\frac{k}{2}(2 + ak(\mathrm{H}_{(ak-2)/4} - \mathrm{H}_{ak/4}))$ |
| exponential | $e^{-(h/a)}$ | $a$ | $\frac{k}{1+ka}$ |
| gaussian | $e^{-(h/a)^2}$ | $a\Gamma(3/2)$ | $ke^{k^2a^2/4}\mathrm{erfc}(ak/2)$ |
| sine | $\frac{1}{h}\sin(h/a)$ | $\frac{\pi}{2}$ | $\frac{2k}{\pi}\mathrm{ArcCot}(ak)$ |
| Liouville | $h^{a-1}$ | $\frac{1}{\Gamma(a)}$ | $k^{1-a}$ |
| constant | $\theta(a-h)$ | $a$ | $\frac{1-e^{-ka}}{a}$ |
| linear | $(a-h)\theta(a-h)$ | $\frac{a^2}{2}$ | $\frac{2}{a^2k}(ak - (1-e^{-ka}))$ |
| parabola | $(a-h)^2\theta(a-h)$ | $\frac{a^3}{3}$ | $\frac{6}{a^3k^2}(\frac{1}{2}ak(ak-2) + (1-e^{-ka}))$ |
| Caputo | $h^{a-1}\theta(x-h)$ | $\frac{1}{\Gamma(a)}$ | $k^{1-a}\left(1 - \frac{\Gamma(a,kx)}{\Gamma(a)}\right)$ (not an eigenfunction) |

An application of

- (a) the local version of the operator $\hat{O}_{\text{local}}$
- (b) a shift operation $\hat{s}$ and
- (c) an averaging procedure $\hat{A}$ with an appropriately chosen weight function $w(a,h)$

$$\langle f(a)\rangle = \hat{A}(a,h)\,f(h) = \frac{1}{N}\int_0^\infty dh\,w(a,h)\,f(h) \tag{8.17}$$

For these three operations there exist six optional permutations, which result in six equally reasonable realizations of nonlocality:

A first set with $\hat{A}\,\hat{s}$ sequence:

$$_1\hat{O}_{\text{nonlocal}}(x)\,f(x) = \hat{O}_{\text{local}}\,\hat{A}\,\hat{s}f(x) \tag{8.18}$$

$$= \frac{1}{N}\hat{O}_{\text{local}}\int_0^\infty dh\,w(a,h)\,\hat{s}f(x) \tag{8.19}$$

$$_2\hat{O}_{\text{nonlocal}}(x)\,f(x) = \hat{A}\,\hat{O}_{\text{local}}\,\hat{s}f(x) \tag{8.20}$$

$$= \frac{1}{N}\int_0^\infty dh\,w(a,h)\hat{O}_{\text{local}}\,\hat{s}f(x) \tag{8.21}$$

$$_3\hat{O}_{\text{nonlocal}}(x)\,f(x) = \hat{A}\,\hat{s}\,\hat{O}_{\text{local}}f(x) \tag{8.22}$$

$$= \frac{1}{N}\int_0^\infty dh\,w(a,h)\,\hat{s}\hat{O}_{\text{local}}f(x) \tag{8.23}$$

and a second set with $\hat{s}\hat{A}$ sequence:

$$_4\hat{O}_{\text{nonlocal}}(x)\,f(x) = \hat{s}\hat{A}\,\hat{O}_{\text{local}}f(x) \tag{8.24}$$

$$= \frac{1}{N}\hat{s}\int_0^\infty dh\,w(a,h)\,\hat{O}_{\text{local}}f(x) \tag{8.25}$$

$$= \hat{s}\hat{O}_{\text{local}}f(x) \tag{8.26}$$

$$_5\hat{O}_{\text{nonlocal}}(x)\,f(x) = \hat{s}\,\hat{O}_{\text{local}}\,\hat{A}f(x) \tag{8.27}$$

$$= \frac{1}{N}\hat{s}\,\hat{O}_{\text{local}}\int_0^\infty dh\,w(a,h)f(x) \tag{8.28}$$

$$= \hat{s}\,\hat{O}_{\text{local}}f(x) \tag{8.29}$$

$$_6\hat{O}_{\text{nonlocal}}(x)\,f(x) = \hat{O}_{\text{local}}\,\hat{s}\,\hat{A}f(x) \tag{8.30}$$

$$= \frac{1}{N}\hat{O}_{\text{local}}\,\hat{s}\int_0^\infty dh\,w(a,h)f(x) \tag{8.31}$$

$$= \hat{O}_{\text{local}}\,\hat{s}f(x) \tag{8.32}$$

In the presentation of the last three realizations of a nonlocal operator we made use of the normalizability of the weight-function. These realizations may be interpreted as simple models of a delayed response of a physical system.

For example for the local differential equation

$$\frac{d}{dt}y(t) = -ky(t) \tag{8.33}$$

the nonlocal pendant is given according to (8.24) and (8.27) as

$$\hat{s}\,\frac{d}{dt}y(t) = -ky(t) \tag{8.34}$$

A multiplication from the left with the inverse of the shift operator (e.g. for the case $\hat{s}_{\pm}^{-1} = \hat{s}_{\mp}$) leads to

$$\hat{s}^{-1}\hat{s}\,\frac{d}{dt}y(t) = -k\hat{s}^{-1}y(t) \tag{8.35}$$

$$\frac{d}{dt}y(t) = -ky(t \mp h) \tag{8.36}$$

A similar treatment for (8.30) leads to

$$\frac{d}{dt}\hat{s}\,y(t) = -ky(t) \tag{8.37}$$

$$\frac{d}{dt}y(t \pm h) = -ky(t) \tag{8.38}$$

Such a behaviour could be called discontinuous nonlocality. A promising field of application might be in signal theory, where the parameter $h$ determines a fixed time delay for a reaction of a given system [May(1975)].

The first two realizations of a nonlocal operator (8.18), (8.20) are equivalent which follows from:

$$[\hat{O}_{\text{local}}(x), \hat{A}(a,h)] = 0 \tag{8.39}$$

In other words, the weight procedure $\hat{A}$ and the local operator $\hat{O}$ commute, as long as the weight procedure is independent of $x$.

The equivalence ${}_2\hat{O}_{\text{nonlocal}}(x) = {}_3\hat{O}_{\text{nonlocal}}(x)$ follows, as long as the local operator is a pure derivative, e.g. $\hat{O}_{\text{local}} = d/dx$, from

$$[\hat{O}_{\text{local}}(x), \hat{s}] = 0 \tag{8.40}$$

or in other words, as long as $\hat{O}_{\text{nonlocal}}(x)$ commutes with the shift operator $\hat{s}$, which is true for a pure derivative operator.

As a first intermediate result we may conclude, that our presentation of a nonlocal generalization of a local operator in the special case of the local derivative leads to only one reasonable definition which according to (8.18) is given by:

$$\frac{d}{dx}_{\text{nonlocal}} f(x) = \frac{1}{N} \frac{d}{dx} \int_0^\infty dh\, w(a,h)\, \hat{s} f(x) \tag{8.41}$$

This definition allows for a generalization of simple differential equations.

As an example, the first order differential equation

$$\frac{d}{dx} f(x) = \kappa f(x) \tag{8.42}$$

may be extended to the causal nonlocal case

$$\frac{d}{dx}_{\text{nonlocal}} f(x) = \frac{1}{N} \frac{d}{dx} \int_0^\infty dh\, w(a,h)\, f(x-h) = \kappa f(x) \tag{8.43}$$

With the ansatz

$$f(x) = e^{kx} \tag{8.44}$$

we obtain

$$\frac{d}{dx}_{\text{nonlocal}} e^{kx} = \frac{1}{N} \frac{d}{dx} \int_0^\infty dh\, w(a,h)\, e^{k(x-h)} \tag{8.45}$$

$$= \frac{1}{N} \int_0^\infty dh\, w(a,h)\, k e^{k(x-h)} \tag{8.46}$$

$$= e^{kx} \frac{1}{N} \int_0^\infty dh\, w(a,h)\, k e^{-kh} \tag{8.47}$$

Therefore the exponential function (8.44) indeed is an eigenfunction of the nonlocal differential equation (8.43) and the eigenvalue spectrum is given by

$$\kappa(k) = k \frac{1}{N} \int_0^\infty dh\, w(a,h)\, e^{-kh} \tag{8.48}$$

In Table 8.1 this spectrum is calculated analytically for different weight functions and the corresponding graphs are given in Fig. 8.3. Despite the fact, that the analytic solutions cover a wide range of exotic functions for different weights, the graphs differ not very much and show a similar behaviour: For finite values of $a$ the eigenvalue spectrum increases weaker than $\kappa(k) = k$ for the local limit $a \to 0$.

At this point of discussion we may now try to determine the connection between nonlocal operators and fractional calculus. For that purpose, we will weaken our requirements for the weight function and will first give up the requirement of normalizability (8.11). In addition, we allow for weakly singular weight functions. In that case, we may consider:

$$w(a,h) = h^{a-1}, \quad 0 < a < 1 \tag{8.49}$$

$$N = \Gamma(a) \tag{8.50}$$

With these settings (8.41) is nothing else but the Liouville definition of a fractional derivative. To be more specific, depending on our choice of the shift operator, the four cases (8.12) lead to

$$_1\hat{O}(\hat{s}_-)\, f(x) = \frac{1}{\Gamma(a)} \frac{d}{dx} \int_0^\infty dh\, h^{a-1} f(x-h) \tag{8.51}$$

$$= \frac{1-a}{\Gamma(a)} \int_0^\infty dh\, h^{a-1} \frac{f(x) - f(x-h)}{h} \tag{8.52}$$

$$_1\hat{O}(\hat{s}_+)\, f(x) = \frac{1}{\Gamma(a)} \frac{d}{dx} \int_0^\infty dh\, h^{a-1} f(x+h) \tag{8.53}$$

$$= \frac{1-a}{\Gamma(a)} \int_0^\infty dh\, h^{a-1} \frac{f(x+h) - f(x)}{h} \tag{8.54}$$

$$_1\hat{O}\left(\frac{\hat{s}_+ + \hat{s}_-}{2}\right) f(x) = \frac{1}{\Gamma(a)} \frac{d}{dx} \int_0^\infty dh\, h^{a-1} \frac{f(x+h) + f(x-h)}{2} \tag{8.55}$$

$$= \frac{1-a}{\Gamma(a)} \int_0^\infty dh\, h^{a-1} \frac{f(x+h) - f(x-h)}{2h} \tag{8.56}$$

$$_1\hat{O}\left(\frac{\hat{s}_+ - \hat{s}_-}{2h}\right) f(x) = \frac{1}{\Gamma(a)} \frac{d}{dx} \int_0^\infty dh\, h^{a-1} \frac{f(x+h) - f(x-h)}{2h} \tag{8.57}$$

$$= \frac{2-a}{2\Gamma(a)} \int_0^\infty dh\, h^{a-1} \frac{f(x+h) - 2f(x) + f(x-h)}{h^2} \tag{8.58}$$

which correspond to the left-Liouville and right-Liouville (5.57), Feller ${}_{\mathrm{F}}D_1^{\alpha}(\theta = 1)$ (5.79) and Riesz (5.71) definition of a fractional derivative. Hence these four different definitions of a fractional derivative operator may be understood within the framework of a nonlocal theory as a result of left, right, symmetric and anti-symmetric shift operations.

In addition we want to emphasize the smooth transition from normalizable, nonsingular weights to non-normalizable, weakly-singular weight functions, as long as the eigenvalue spectrum for a nonlocal differential equation is considered. In Fig. 8.3 we have plotted the eigenvalues of the solutions of the first order differential equation (8.48) for the left-sided Liouville fractional derivative, which are simply given as $\kappa(k) = k^{1-a}$. For $k > 1$ they show a similar behaviour as the nonlocal solutions using a normalizable weight.

The fractional derivative definition may be considered as a convolution integral with a weakly singular kernel [Gorenflo(2008)] and the concept of a fractional derivative may be embedded as a special case into a general theory of nonlocal operators.

The presented theory of nonlocal operators is not restricted to pure derivative operators but may be applied to any coordinate dependent local operator. If we intend to apply the fractional calculus to arbitrary curvilinear coordinate systems, the concept of fractional calculus has to be extended to arbitrary space dependent operators.

It should be emphasized, that space dependent local operators will not commute with the shift operator in general:

$$[\hat{O}_{\text{local}}(x), \hat{s}] \neq 0 \tag{8.59}$$

We therefore have two equivalent extensions of a fractional generalization of an arbitrary operator, namely $\hat{O}_1$ and $\hat{O}_3$ which according to (8.18) and (8.22) are given by:

$$\hat{O}_{LR}\, f(x) = \frac{1}{\Gamma(a)} \hat{O}_{\text{local}}(x) \int_0^\infty dh\, h^{a-1} \hat{s} f(x) = \hat{O}^a\, f(x) \tag{8.60}$$

We call $\hat{O}_{LR}$ as the Liouville–Riemann or covariant $\hat{O}^a$ fractional extension for an arbitrary coordinate dependent local operator $\hat{O}$ and

$$\hat{O}_{LC}\, f(x) = \frac{1}{\Gamma(a)} \int_0^\infty dh\, h^{a-1} \hat{s}\, \hat{O}_{\text{local}}(x)\, f(x) = \hat{O}_a\, f(x) \tag{8.61}$$

we call $\hat{O}_{LC}$ as the Liouville–Caputo or $\hat{O}_a$ contravariant fractional extension for an arbitrary coordinate dependent local operator $\hat{O}$.

With (8.60) and (8.61) we have expanded the concept of fractional calculus from an extension of the standard derivative operator $d/dx$ to a specific nonlocal generalization of an arbitrary local operator.

Within the framework of nonlocal operator calculus we may now discuss problems like coordinate transformations or an extension to higher-dimensional Riemannian spaces from a general point of view. For example an explicit and complete derivation of fractional operators in spherical coordinates is still an open task, despite the fact, that there have been several attempts in the past [Goldfain(2006), Tarasov(2008), Roberts(2009), Li(2010)].

Besides a geometric interpretation of the fractional integral in terms of a given area of projection of a curved fence presented in Chapter 5.6 proposed by Podlubny [Podlubny(2002)], in this chapter we gave an alternative interpretation in terms of an averaging weight procedure for a given function.

This is in close analogy to possible interpretations of the standard integral, which may be interpreted geometrically as the calculation of an area under a given curve and a function average in a given interval respectively.

## 8.3 Memory effects

The solutions of a differential equation may be given in general, but for a specific application they have to obey additional initial or boundary conditions. If the differential equation is defined in terms of local operators, e.g. the Newtonian equations of motion, the knowledge of an initial position and the initial velocity determines a solution completely. The solutions of the general wave equation for a vibrating string may be determined, assuming the string is fixed at both ends. In both cases a specific solution of a local differential equation is completely determined by local conditions. As a consequence, more complex problems may be solved by piecewise solution in different areas, which later are combined to obtain a smooth general solution.

Typical examples are scattering problems. The Mie solutions of the Maxwell equations [Mie(1908)] e.g. describe the scattering of plane electromagnetic waves with spheres. For that purpose, the incoming wave, the scattered wave and the wave inside the sphere are expanded in spherical vector functions, which at first are independent of each other, but are combined employing the boundary conditions given on the surface of the sphere.

Another example is the penetration of a double humped fission barrier [Cramer(1970)]. The complex double humped potential is separated into different regions of space, where the potential takes the simple form of a harmonic oscillator. The Schrödinger equation is solved in these different regions and finally the different solutions are combined to obtain a smooth, differentiable general solution.

All these strategies are well established techniques to obtain solutions for a local differential equation. But they all will fail if applied to nonlocal theories.

The reason of the failure of the above mentioned techniques is of course the nonlocality of the operators used. The averaging procedure $\hat{A}$ spreads out over the full region of space.

In the following we will demonstrate, how nonlocality influences the behaviour of solutions and how the actual state of a system is influenced by its history.

We will demonstrate this memory effect investigating the properties of an idealized model for a battery, which is discharged via a resistor [Westerlund(1991)]. First we present the classical local solution. Then we will solve the corresponding nonlocal differential equation and investigate the implicit and explicit influence of memory.

- In a classical approach, the discharging of a battery may simply be described as an RC-circuit with resistance R and capacity C, which yields a local first order differential equation for the charge $Q$:

$$\frac{d}{dt}Q(t) = -\frac{1}{RC}Q(t), \qquad t \geq 0 \tag{8.62}$$

with the ansatz $Q(t) = Q_0 e^{-kt}$ a solution is given by:

$$k = \frac{1}{RC} \tag{8.63}$$

$$Q(t = 0) = Q_0 \tag{8.64}$$

where $k$ is called the decay constant and $Q_0$ is the charge amount for the battery at $t = 0$, which is given by $Q_0 = U_0 C$, where $U_0$ is the battery (capacitor) voltage at $t = 0$.

This is the classical description of the process. There is no information necessary on the number of charge-cycles or the age of the battery. In contrast to this idealized result, in practice a lowered capacity after repeated charge/discharge cycles has been reported for alcaline batteries [Hullmeine(1989), Tarascon(1996), Sato(2001)]. We want to incorporate such memory effects into our idealized local model. Hence a promising strategy might be the step from a local to a nonlocal description:

- In the next step we will solve the corresponding nonlocal differential equation:

$$\frac{d}{dt}_{\text{nonlocal}} Q(t) = -\frac{1}{RC}Q(t), \qquad t \geq 0 \tag{8.65}$$

With (8.41) we obtain using the causal translation operator $\hat{s}_-$

$$\frac{d}{dt}_{\text{nonlocal}} Q(t) = \frac{1}{N}\frac{d}{dt}\int_0^\infty dh\, w(a,h)\, Q(t-h) = -\frac{1}{RC}Q(t) \tag{8.66}$$

We will use the following weight-function

$$w(a,t) = e^{-t/a} \tag{8.67}$$

$$N = a \tag{8.68}$$

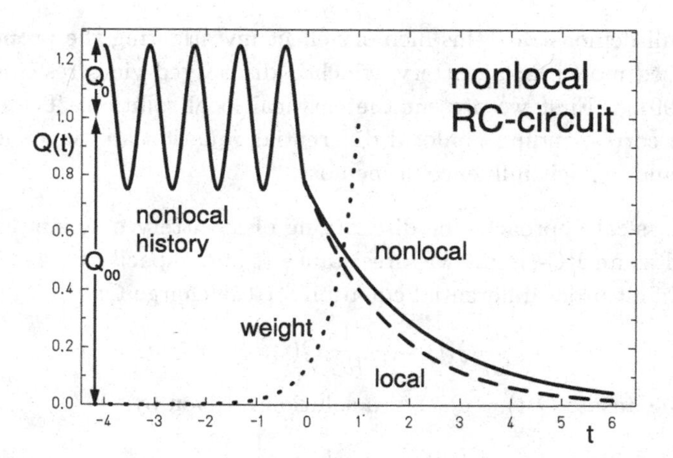


Fig. 8.4. Solutions of the nonlocal differential equation (8.73). Thick lines indicate the non-local solution (8.72), dashed line denotes the corresponding local solution ($a = 0$). The dotted line is a plot of the weight function $w(a,t)$ at $t = 1$ to give an impression about the influence of the memory effect.

which has the advantage, that all following results are obtained analytically. Inserting the ansatz

$$Q(t) = Q_0 e^{-kt} \qquad -\infty \leq t \leq +\infty \tag{8.69}$$

into (8.66) we obtain the conditions

$$k = \frac{1}{RC + a} \tag{8.70}$$

$$Q(0) = Q_0 \tag{8.71}$$

The decay constant $k$ for a given fixed set $Q_0, R, C$ is reduced compared to the local case.

Despite the fact, that the differential equation is valid for $t > 0$ only, we had to give additional information on $Q(t)$ in the range $-\infty < t < 0$ in order to evaluate the integral (8.66). The eigenfunction (8.69) makes an implicit assumption about the development of the solution in the past.

But $Q(t) = Q_0 e^{-kt}$ is not a very realistic history, especially for $t \ll 0$.

- Let us therefore take into account a more realistic charge/discharge history of the battery. We will try

$$Q(t) = \begin{cases} Q_{00} + Q_0 \cos(\omega t) & t < 0 \\ (Q_{00} + Q_0) e^{-kt} & t \geq 0 \end{cases} \tag{8.72}$$

which is an idealized model for a charge history for $t < 0$, where $Q_{00}$ is the average over all charge cycles, $Q_0$ is the amplitude of a harmonic oscillating charge/discharge cycle and $\omega$ is the cycle frequency.
Inserting this ansatz into (8.66) we obtain

$$\frac{d}{dt}_{\text{nonlocal}} Q(t) = \frac{1}{a}\frac{d}{dt}\int_0^\infty dh\, e^{-h/a}\, Q(t) \tag{8.73}$$

$$= \frac{1}{a}\frac{d}{dt}\int_0^t dh\, e^{-h/a}\,(Q_{00}+Q_0)e^{-k(t-h)}$$

$$+\frac{1}{a}\frac{d}{dt}\int_t^\infty dh\, e^{-h/a}\,(Q_{00}+Q_0\cos(\omega(t-h))) \tag{8.74}$$

$$= (Q_{00}+Q_0)\left(e^{-kt}\frac{k}{-1+ak} + e^{-t/a}\frac{1}{a-a^2k}\right)$$

$$-e^{-t/a}\frac{Q_0+Q_{00}(1+a^2\omega^2)}{a+a^3\omega^2} \tag{8.75}$$

$$= -\frac{1}{RC}Q(t), \qquad t \geq 0 \tag{8.76}$$

A term by term comparison leads to the conditions

$$e^{-kt}\frac{k}{-1+ak} = -\kappa e^{-kt} \tag{8.77}$$

$$e^{-t/a}\left(\frac{Q_0+Q_{00}}{a-a^2k} - \frac{Q_0+Q_{00}(1+a^2\omega^2)}{a+a^3\omega^2}\right) = 0 \tag{8.78}$$

From the first condition we obtain

$$k = \frac{1}{RC+a} \tag{8.79}$$

which coincides with (8.70).

For the solution of the local differential equation the behaviour for $t < 0$ was arbitrary. Now for the nonlocal case we derived two solutions which are identical for $t \geq 0$, but differ in the past $t < 0$.

The second condition (8.78) yields:

$$\frac{1}{RC} = \frac{-Q_0}{Q_0+Q_{00}}\frac{a\omega^2}{1+a\omega^2} \tag{8.80}$$

from which we deduce, that an exponential discharge is only possible for $Q_0 < 0$. This implies, that Q(0) is the minimum value of the oscillating part. Hence we have found only a special solution for a given set of $a, Q_{00}, Q_0, \omega$ values. For value sets, which do not fulfil condition (8.80) therefore the solution is not simply an exponential function any more, but of a rather more general function type.

We may conclude, that a memory effect may be modelled with nonlocal differential equations. We presented an idealized model to describe the behaviour of a discharged battery. Indeed we found, that the discharge speed is reduced, if a memory effect ($a > 0$) is present. Consequently this simple model may be useful in biophysics, since it demonstrates, that training has a positive impact on fatigue behaviour.

The most surprising news is, that the knowledge of the full past ($-\infty < t \leq 0$) of a given system does not uniquely determine the future development. As we have just demonstrated, an oscillatory and an exponential decay behaviour respectively both may lead to the same exponential decay rate in the future. As a consequence, a given behaviour in the future $t > 0$ may be the result of different histories. This is just the inverted concept of bifurcation theory, which states, that despite an identical behaviour in the past, a minimal parameter change may cause a drastic change in the future development of a system.

We may easily understand the reason for this behaviour: A nonlocal operator is not determined by the function values directly; by definition only a weighted integral over all past function values is necessary. Of course this may yield the same integral value at e.g. $t = 0$ for different functions. Therefore the concept of nonlocality, in analogy to Heisenberg's uncertainty principle, introduces a new quality of uncertainty into deterministic physical systems.

It should also be clear now, that the method of a piecewise solution will not be applicable in general for nonlocal problems [Jeng(2008)], but will only work for very special scenarios.

In addition a covariant formulation of a fractional derivative is necessarily based on a global concept of nonlocality, which is valid for space dependent derivative operators too.

## Chapter 9

# Quantum Mechanics

*So far we have presented several methods to extend the definition of a standard derivative from integer value n to arbitrary $\alpha$ values:*

$$\frac{d^n}{dx^n} \to \frac{d^\alpha}{dx^\alpha} \tag{9.1}$$

*Indeed we may argue, that nearly all fractional extensions of classical problems are based on this ansatz. But of course, the proposed fractional extensions of a derivative operator leave the role of the corresponding coordinates untouched. This is one of the reasons of the difficulties to determine the fractional extension of a derivative e.g. in spherical coordinates and one of the reasons that nearly all the results of publications are derived in Cartesian coordinates.*

*A simple Fourier-transform of (9.1) using (5.8) results in*

$$\frac{d^n}{dx^n} \to k^n \tag{9.2}$$

$$\frac{d^\alpha}{dx^\alpha} \to k^\alpha \tag{9.3}$$

*and leads to the conclusion, that in the frequency domain*

$$k^n \to k^\alpha \tag{9.4}$$

*is an equivalent formulation of fractional calculus.*

*It is indeed remarkable, that many problems of fractional calculus may be easily solved applying a simple Fourier transform or alternatively for initial value problems by a Laplace-transformation.*

*On the other hand, any theory which works with fractional derivatives only and neglects an adequate treatment of coordinates, may be transformed into an equivalent standard differential theory with fractional powers*

$$\left\{x^n, \frac{d^\alpha}{dx^\alpha}\right\} \equiv \left\{\frac{d^n}{dk^n}, k^\alpha\right\} \tag{9.5}$$

*and consequently lead to results, which may be derived with standard methods without knowledge of fractional calculus.*

*A possible realization of this concept is the transition from integer space dimension $n$ to arbitrary space dimension $\alpha$. A geometrical interpretation could be given in terms of the volume of a hypercube in such a space.*

*If we reduce the concept of fractal geometry to the introduction of arbitrary space dimensions, which of course is an oversimplification, fractional calculus and fractal geometry are indeed convertible via a simple Fourier transform and therefore are equivalent. Some aspects about a relationship between fractional calculus and fractals are discussed in [Nigmatullin(1992), Tatom(1995)].*

*The transition from classical mechanics to quantum mechanics may be interpreted as a transition from independent coordinate space and momentum space to a Hilbert space, in which space and momentum operators are treated similarly. Consequently one postulate of quantum mechanics states, that derived results must be independent of the specific choice of e.g. a space or momentum representation.*

*This implies, that a successful fractional extension of quantum mechanics has to treat coordinates and conjugated momenta equivalently:*

$$\left\{x^n, \frac{d^n}{dx^n}\right\} \rightarrow \left\{x^\alpha, \frac{d^\alpha}{dx^\alpha}\right\} \tag{9.6}$$

*This simultaneous treatment is the major difference between a classical and a quantum mechanical treatment. Furthermore this is the key information needed to quantize any classical quantity like a Hamilton function or angular momentum.*

*In the following we will apply the method of canonical quantization to the classical Hamilton function to obtain a fractional extension of the standard Schrödinger equation and will investigate some of its properties.*

*One surprising result will be the observation, that confinement is a natural outcome of the use of a fractional wave equation.*

*Especially the transformation properties under rotations will lead to the conclusion, that the behaviour of a fractional Schrödinger equation should be directly compared to a standard Pauli equation.*

*A fractional Schrödinger equation describes particles with a complex internal structure, which we call fractional spin.*

## 9.1 Canonical quantization

The transition from classical mechanics to quantum mechanics may be performed by canonical quantisation [Dirac(1930), Messiah(1968)]. The classical canonically conjugated observables $x$ and $p$ are replaced by quantum mechanical observables $\hat{x}$ and $\hat{p}$, which are introduced as derivative operators on a Hilbert space of square integrable wave functions $f$. The space coordinate representations of these operators are:

$$\hat{x}f(x) = xf(x) \tag{9.7}$$

$$\hat{p}f(x) = -i\hbar\partial_x f(x) \tag{9.8}$$

where $\hat{x}$ and $\hat{p}$ fulfil the commutation relation

$$[\hat{x}, \hat{p}] = i\hbar \tag{9.9}$$

We will now describe a generalisation of these operators from integer order derivative to arbitrary order derivative.

According to our comments on parity conservation (7.14) we introduce two canonically conjugated operators in space representation:

$$\hat{X}\, f(\hat{x}^\alpha) = \left(\frac{\hbar}{mc}\right)^{(1-\alpha)} \hat{x}^\alpha\, f(\hat{x}^\alpha) \tag{9.10}$$

$$\hat{P}\, f(\hat{x}^\alpha) = -i\left(\frac{\hbar}{mc}\right)^{\alpha} mc\, \hat{D}^\alpha\, f(\hat{x}^\alpha) \tag{9.11}$$

The attached factors $(\hbar/mc)^{(1-\alpha)}$ and $(\hbar/mc)^\alpha mc$ ensure correct length and momentum units. For the special case $\alpha = 1$ these definitions correspond to the classical limits (9.7) and (9.8). The Hilbert space of square integrable functions $f, g$ is based on the scalar product

$$< f|g >= \int dx f^* g \tag{9.12}$$

Expectation values of an operator $\hat{O}$ may be calculated with

$$< f|\hat{O}|g >= \int dx f^* \hat{O} g \tag{9.13}$$

## 9.2 Quantization of the classical Hamilton function and free solutions

With definitions (9.10), (9.11) for the space and momentum operator we may now quantize the classical Hamilton function of a non-relativistic particle.

The classical non-relativistic $N$ particle Hamilton function $H_c$ depends on the classical moments and coordinates $\{p_i, x^i\}$:

$$H_c = \sum_{i=1}^{3N} \frac{p_i^2}{2m} + V(x^1, ..., x^i, ..., x^{3N}) \tag{9.14}$$

Following the canonical quantization method the classical observables are replaced by quantum mechanical operators. Hence we obtain the fractional Hamiltonian $H^\alpha$:

$$H^\alpha = \sum_{i=1}^{3N} \frac{\hat{P}_i^2}{2m} + V(\hat{X}^1, ..., \hat{X}^i, ..., \hat{X}^{3N}) \tag{9.15}$$

$$= -\frac{1}{2}mc^2 \left(\frac{\hbar}{mc}\right)^{2\alpha} \hat{D}^{\alpha i}\hat{D}_i^\alpha + V(\hat{X}^1, ..., \hat{X}^i, ..., \hat{X}^{3N}) \tag{9.16}$$

Thus, a time dependent Schrödinger type equation for fractional derivative operators results

$$H^\alpha \Psi = i\hbar \frac{\partial}{\partial t}\Psi \tag{9.17}$$

$$= \left(-\frac{1}{2}mc^2 \left(\frac{\hbar}{mc}\right)^{2\alpha} \hat{D}^{\alpha i}\hat{D}_i^\alpha + V(\hat{X}^1, ..., \hat{X}^i, ..., \hat{X}^{3N})\right)\Psi$$

where we have introduced Einstein's summation convention $\sum_{i=1}^N x_i^2 = x^i x_i$ For $\alpha = 1$ this reduces to the classical Schrödinger equation.

A stationary Schrödinger equation results with the product ansatz $\Psi(\hat{X}^i, t) = \psi(\hat{X}^i)T(t)$ and a separation constant $E$:

$$H^\alpha \psi(\hat{X}^i) = \left(-\frac{1}{2}mc^2 \left(\frac{\hbar}{mc}\right)^{2\alpha} \hat{D}^{\alpha i}\hat{D}_i^\alpha + V(\hat{X}^1, ..., \hat{X}^i, ..., \hat{X}^{3N})\right)\psi(\hat{X}^i)$$

$$= E\psi(\hat{X}^i) \tag{9.18}$$

We will investigate the simple case of a free one dimensional fractional Schrödinger equation:

$$H_{\text{free}}^\alpha \psi = -\frac{1}{2}mc^2 \left(\frac{\hbar}{mc}\right)^{2\alpha} \hat{D}^\alpha \hat{D}^\alpha \psi = E\psi \tag{9.19}$$

This fractional differential equation is equivalent to the fractional regularized wave equation (7.21).

For $\alpha = 1$ solutions are given as plane waves:

$$\psi^+(x) = \cos(kx) \tag{9.20}$$

$$\psi^-(x) = \sin(kx) \tag{9.21}$$

If we apply the fractional derivative definition according to Fourier, the solutions of the free fractional Schrödinger equation are given by (7.22), for the Riemann definition by (7.24) and for the Caputo definition by (7.26).

In addition a valid solution of the fractional Schrödinger equation must be normalizable, so that a probabilistic interpretation is possible. In the region $\alpha < 1$ this may be realized easily:

For the Fourier fractional derivative the norm may be calculated analytically:

$$\int_{-\infty}^{\infty} dx_{\mathrm{F}} \sin(\alpha, x)_{\mathrm{F}} \sin(\alpha, x) = \frac{(-1 + \cos(\frac{\pi}{\alpha}))}{4\pi \cos(\frac{\pi}{2\alpha})} \tag{9.22}$$

$$\int_{-\infty}^{\infty} dx_{\mathrm{F}} \sin(\alpha, x)_{\mathrm{F}} \cos(\alpha, x) = 0 \tag{9.23}$$

$$\int_{-\infty}^{\infty} dx_{\mathrm{F}} \cos(\alpha, x)_{\mathrm{F}} \cos(\alpha, x) = -\frac{(3 + \cos(\frac{\pi}{\alpha}))}{4\pi \cos(\frac{\pi}{2\alpha})} \tag{9.24}$$

For $\alpha \geq 1$ this integral is divergent. In analogy to classical quantum mechanics we hence propose a box normalization in this region, where the dimensions of the box should be chosen large enough.

A similar procedure is necessary for the solutions according to Riemann and Caputo. In this case the solutions correspond to solutions for a Schrödinger equation in an infinitely deep potential well. This is equivalent to a search for zeroes for these solutions and may be easily performed numerically.

A graphical representation of the solutions of the free fractional Schrödinger equation is given in Fig. 9.1.

For a physical interpretation it is important to recognize, that the solutions for $\alpha < 1$ are more and more located at the origin.

This is in contrast to the solutions for free particles, which are described by the ordinary Schrödinger equation as free waves which are spread out on the whole space and therefore are located everywhere with a similar probability (this is a direct consequence of Heisenberg's uncertainty relation, since the momentum of the free particle is known exactly, we may not make a statement of the position).

If Bohr's concept of a probability interpretation is valid for the solution of the fractional Schrödinger equation for a fractional particle we obtain the remarkable result, that a fractional particle will not be equally spaced, but instead is located at the origin.

As a first reaction, we could call this behaviour non-physical and might discard a fractional Schrödinger equation. On the other hand we know

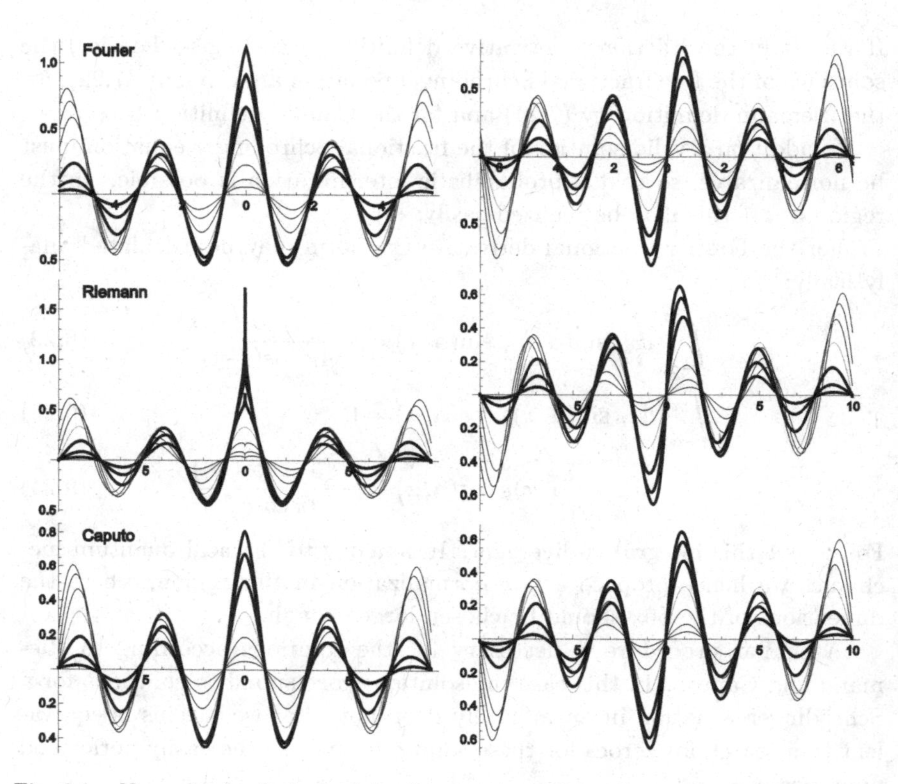


Fig. 9.1. Normalized solutions of the free fractional Schrödinger equation with good parity (even on the left and odd on the right side) for different fractional derivative parameters (thick lines $\alpha < 1$ and thin lines $\alpha \geq 1$) according to Fourier, Riemann and Caputo: Solutions in the region $\alpha \geq 1$ are box normalizable. Solutions for $\alpha < 1$ are localized at $x = 0$ and decrease very fast for large $x$-values.

about particles, which undoubtedly exist but have not been observed as free particles until now. These particles are called quarks and the problem, that they have not been observed as free quarks until now, is called quark confinement and is a direct consequence of the specific nature of strong interaction.

Hence, we will not discard the fractional Schrödinger equation, but will compare our predictions with results which have been derived by using traditional methods for a description of hadronic interaction.

Of course it would be a nice success of fractional calculus, if we could derive a fractional theory of quark dynamics, which was more elegant and successful than the usually applied methods.

A phenomenological strategy to model quark confinement is a classical description in terms of an ad hoc postulated potential. The most prominent example is a linear ansatz [Eichten(1975)]:

$$V(x) \sim \text{const}\,|x| \tag{9.25}$$

In the next section we want to find out, whether we can reproduce the behaviour of the free solutions of the fractional Schrödinger equation (9.19) if we assume, that similar solutions of the standard ($\alpha = 1$) Schrödinger equation with an additional phenomenological potential term exist.

$$H^{\alpha}_{\text{free}}\psi = E\psi \approx \left(-\frac{\hbar^2}{2m}\partial_x^2 + V(x)\right)\psi \tag{9.26}$$

We will show, that the explicit form of such an additional potential term may indeed be derived, using methods, originally developed to describe the fission yields of excited nuclei for a given potential energy surface.

## 9.3 Temperature dependence of a fission yield and determination of the corresponding fission potential

We will investigate the problem, whether we can estimate the original form of a given potential, if a complete set of eigenfunctions and corresponding energy values of the standard Schrödinger equation with a potential term are known.

For that purpose, we will use methods developed within the framework of fragmentation theory [Fink(1974), Sandulescu(1976), Maruhn(1980), Greiner(1995)], which originally were used for an explanation of the fission yields of heavy nuclei [Lustig(1980)]. We will show, that the temperature dependence of a fission yield and the corresponding potential are indeed connected. For reasons of simplicity, we will present the analytically solvable example of the harmonic oscillator.

The stationary solutions of the Schrödinger equation

$$\left(-\frac{\hbar^2}{2m}\partial_x^2 + V(x)\right)\psi = E\psi \tag{9.27}$$

with the model potential of the harmonic oscillator

$$V(x) = \frac{1}{2}m\omega^2 x^2 \tag{9.28}$$

using the abbreviation

$$\beta = \sqrt{\frac{m\omega}{\hbar}} \tag{9.29}$$

lead to a set of eigenfunctions ($H_n$ denotes the Hermite polynomial)

$$\psi_n(x) = \sqrt{\frac{\beta}{\pi^{\frac{1}{2}} 2^n n!}} H_n(\beta x) e^{-\frac{1}{2}\beta^2 x^2} \tag{9.30}$$

and eigenvalues:

$$E_n = \hbar\omega(n + \frac{1}{2}) \tag{9.31}$$

The temperature dependent fission yield is given by [Maruhn(1980)]:

$$Y(x,T) = \frac{\sum_n \psi_n(x)\psi_n^*(x) e^{-\frac{E_n}{T}}}{\sum_n e^{-\frac{E_n}{T}}} \tag{9.32}$$

Inserting energies (9.31) and eigenfunctions (9.30) using the abbreviation

$$q = e^{-\frac{\hbar\omega}{T}} \tag{9.33}$$

lead to[1]:

$$\begin{aligned} Y(x,T) &= (1-q)\frac{\beta}{\pi^{\frac{1}{2}}} e^{-\beta^2 x^2} \sum_n H_n^2(\beta x) \frac{q^n}{2^n n!} \\ &= (1-q)\frac{\beta}{\pi^{\frac{1}{2}}} e^{-\beta^2 x^2} \frac{1}{\sqrt{1-q^2}} e^{-\frac{2\beta^2 x^2 q}{1+q}} \end{aligned} \tag{9.34}$$

For the limiting case of high temperatures

$$q \approx 1 - \frac{\hbar\omega}{T} \tag{9.35}$$

we obtain:

$$Y(x,T) = \frac{\beta}{\pi^{\frac{1}{2}}} \sqrt{\frac{\hbar\omega}{2T}} e^{-\frac{m\omega^2 x^2}{2T}} \tag{9.36}$$

or

$$Y(x,T) = \sqrt{\frac{m\omega^2}{2\pi T}} e^{-\frac{V}{T}} \tag{9.37}$$

We obtain the important result, that we may deduce the general form of the potential energy $V(x)$ from a temperature dependent fission yield using (9.37). Once eigenfunctions and energy levels are known, we may determine this potential.

[1] A nice derivation is presented in: Greiner, W., Neise, L. and Stöcker, H. (1995) *Thermodynamics and statistical mechanics*, Springer, New York, USA. Exercise 10.7, pp. 280–284.

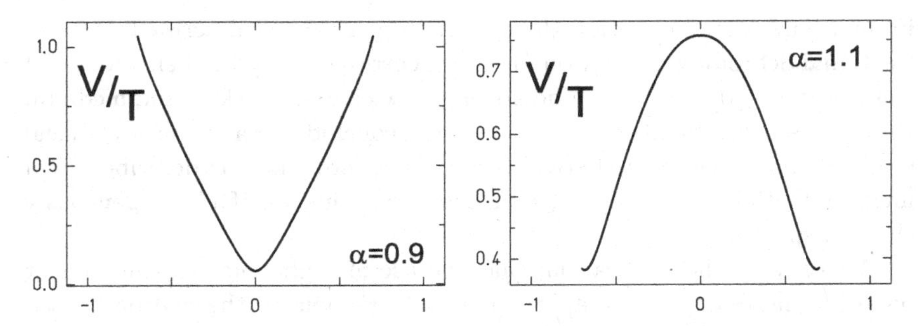


Fig. 9.2. Graph of the potential $V(x)$ which is necessary to simulate the behaviour of eigenfunctions and solutions of the free fractional Schrödinger equation (9.19) in terms of the ordinary Schödinger equation according to (9.26). For $\alpha < 1$ the potential is dominated by a linear term $V \sim \text{const}\,|x|$. For $\alpha > 1$ the potential is of the type $V \sim \text{const}(1 - x^2)$.

With this information we may calculate the regularized solutions of the free fractional Schrödinger equation using the Caputo derivative, normalize these solutions in a box of size $L = 2$ and apply the boundary conditions $\psi(\alpha, x = -1) = 0$ and $\psi(\alpha, x = +1) = 0$. The corresponding eigenvalues are determined numerically.

With these results in a next step we calculate the fission yield (9.32) and deduce the potential according to (9.37). The result is presented in Fig. 9.2.

In the vicinity of $\alpha \approx 1$ we distinguish three cases:

$$V(x) = \begin{cases} c_1 + c_2|x| & \alpha < 1 \\ \text{[3}pt\text{]}0 & \alpha = 1 \\ \text{[3}pt\text{]}c_1(1 - c_2x^2) & \alpha > 1 \end{cases} \tag{9.38}$$

- For the case $\alpha < 1$ the behaviour of the solutions of the free fractional Schrödinger equation corresponds to the behaviour of solutions of the standard Schrödinger equation with a dominant linear potential contribution.
- For the case $\alpha = 1$ both free solutions are identical.
- In order to model the behaviour of the fractional solutions in the case $\alpha > 1$ a potential of type $V \sim c_1(1 - c_2x^2)$ has to be inserted into the standard Schrödinger equation. In three dimensional space this potential is identical to the potential inside a homogeneously charged sphere.

From a classical point of view the behaviour of a particle described by a free fractional Schrödinger equation for $\alpha > 1$ corresponds to the behaviour of a classical charged particle in a homogeneously charged background medium.

On the other hand for $\alpha < 1$ we may conclude, that from a classical point of view, there is a charged background too, which is described by a linear potential term instead of a quadratic term in case of a homogeneously charged sphere.

This is a further indication, that the use of a phenomenological linear potential, necessary for an approximate description for the hadron excitation spectrum in classical models, is obsolete in a fractional theory or in other words, this behaviour is already incorporated in a fractional theory.

In order to clarify this result, in the next section we will present the energy spectrum of the fractional Schrödinger equation with an infinite well potential.

## 9.4 The fractional Schrödinger equation with an infinite well potential

The one dimensional infinite well potential is defined as :

$$V(x) = \begin{cases} 0 & |x| \leq a \\ \infty & |x| > a \end{cases} \tag{9.39}$$

From a classical point of view, the solutions of the standard Schrödinger equation with this potential therefore correspond to the solutions of the classical wave equation for a fixed vibrating string. The reason for this equivalence is based on the fact, that the solution of the standard Schrödinger equation for the infinite potential well may be split as

$$\psi(x) = \begin{cases} \psi_0(x) & |x| \leq a \\ 0 & |x| > a \end{cases} \tag{9.40}$$

Therefore the solution in the interior region $|x| \leq a$ is equivalent to the solution of a free wave equation with boundary conditions $\psi_0(\pm a) = 0$.

Such a piecewise solution is valid only in the case of a standard derivative.

- If we apply a fractional derivative in the sense of Liouville or Fourier, according to our statements on non-locality and memory effects in the previous chapter, the behaviour of the wave function in the outside region influences the behaviour in the inside region. Consequently the use of solutions for the fractional wave equation obtained for a vibrating string will be considered as approximate solutions for the fractional infinite well potential, which are valid only for $\alpha \approx 1$.

- On the other hand, for the Riemann- and Caputo fractional derivative definition, the interior solutions $\psi_0(x)$ for the infinite potential well and for the vibrating string are identical. Indeed, according to (5.32) and (5.43) the fractional integral in this case is confined to the region from 0 to $|x|$ and therefore, the functional behaviour inside the infinite potential well is independent of the exterior region. But if the potential well has a finite depth or is not located at the origin, this equivalence of solutions is lost. The translation invariance of the fractional Schrödinger equation is lost for the Riemann- and Caputo fractional derivative definition.

Therefore until now there is no exact analytic solution for the fractional infinite depth potential well, which is valid for all types of fractional derivatives.

A correct solution was indeed given by a free fractional Schrödinger equation (9.19)

$$H^\alpha \psi = -\frac{1}{2}mc^2 \left(\frac{\hbar}{mc}\right)^{2\alpha} \hat{D}^\alpha \hat{D}^\alpha \psi = E\psi \tag{9.41}$$

where $\psi$ must obey the conditions (9.40).

Instead we will use the boundary conditions

$$\psi(-a) = \psi(a) = 0 \tag{9.42}$$

Conditions (9.40) and (9.42) are only equivalent for the infinite square well potential using the Riemann- and Caputo fractional derivative. They may be considered a crude approximation only in case of the Fourier, Liouville or Riesz fractional derivative.

With these boundary conditions we may indeed use the solutions of the free fractional Schrödinger equation, which for example are given approximately for the Fourier fractional derivative according to (7.22) by

$$_{\mathrm{F}}\sin(\alpha, x) = \frac{1}{2i}\left(e^{\frac{1}{2}e^{\frac{i\pi}{2\alpha}}|x|} - e^{\frac{1}{2}e^{-\frac{i\pi}{2\alpha}}|x|}\right) \tag{9.43}$$

$$_{\mathrm{F}}\cos(\alpha, x) = \mathrm{sign}(x)\frac{1}{2}\left(e^{\frac{1}{2}e^{\frac{i\pi}{2\alpha}}|x|} + e^{\frac{1}{2}e^{-\frac{i\pi}{2\alpha}}|x|}\right) \tag{9.44}$$

the zeroes of these solutions are known analytically (7.32):

$$_{\mathrm{F}}x_0(n, \alpha) = \frac{(n+1)}{\sin(\frac{\pi}{2a})}\frac{\pi}{2} \qquad n = 0, 1, 2, 3... \tag{9.45}$$

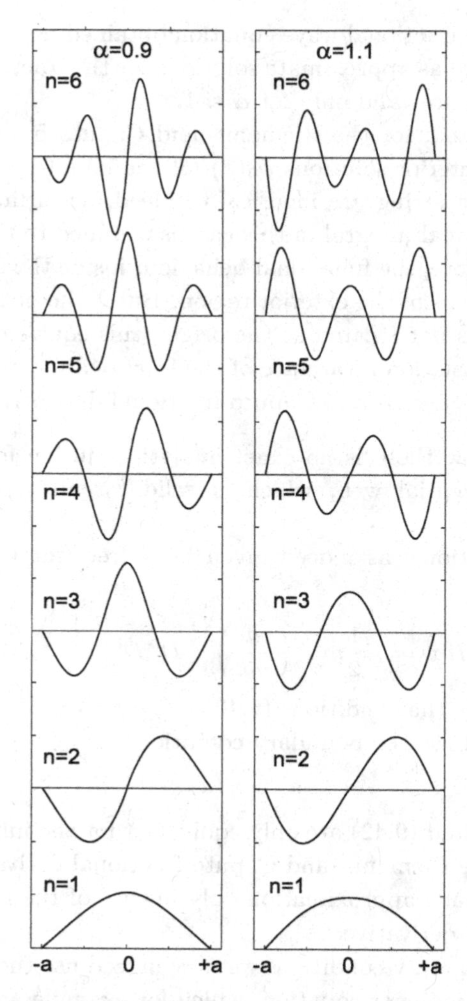


Fig. 9.3. The six lowest eigenfunctions for the one dimensional fractional Schrödinger equation with the infinite square well potential (9.39) based on the Caputo fractional derivative (5.43), with $\alpha = 0.9$ on the left and $\alpha = 1.1$ on the right side.

Therefore approximate solutions of the fractional Schrödinger equation according to Fourier for the infinite depth potential well, which fulfil the boundary conditions (9.42) are given as

$$\psi(\alpha, x) = \begin{cases} {}_{\mathrm{F}}\cos(\alpha, x/x_0(n,\alpha)) & n = 0, 2, 4, \ldots \\ {}_{\mathrm{F}}\sin(\alpha, x/x_0(n,\alpha)) & n = 1, 3, 5, \ldots \end{cases} \tag{9.46}$$

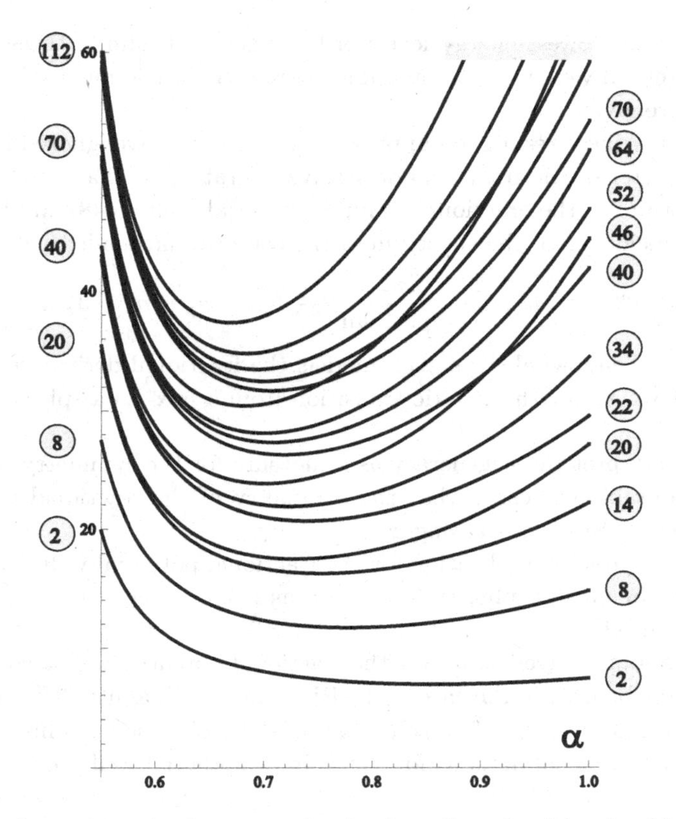


Fig. 9.4. Approximate level spectrum for the three dimensional fractional infinite potential well according to Fourier. Occupation numbers are given in circles. For the semiderivative $\alpha \approx 0.5$ the spectrum more and more coincides with the spectrum of the standard harmonic oscillator.

and the corresponding energy levels follow according to (7.28) as:

$$E(\alpha, n) = \frac{1}{2}mc^2 \left(\frac{\hbar}{mc}\right)^{2\alpha} \left(\frac{\pi}{2\sin(\frac{\pi}{2a})}\right)^{2\alpha} (n+1)^{2\alpha} \tag{9.47}$$

This result may easily be extended to the three dimensional case:

$$E(\alpha, n_x, n_y, n_z) = \frac{1}{2}mc^2 \left(\frac{\hbar}{mc}\right)^{2\alpha} \left(\frac{\pi}{2\sin(\frac{\pi}{2a})}\right)^{2\alpha} \times \left((n_x+1)^{2\alpha} + (n_y+1)^{2\alpha} + (n_z+1)^{2\alpha}\right) \tag{9.48}$$

In Fig. 9.4 the lowest energy levels of the fractional infinite three dimensional potential well are presented for different $\alpha$. Magic numbers for shell closures are given.

Although these results are approximate, since we have ignored the nonlocal character of the fractional derivative operator, we may deduce, that the spectrum of the fractional infinite potential well (9.48) in the limit $\alpha \to \frac{1}{2}$ coincides with the spectrum of the classical harmonic oscillator:

$$\lim_{\alpha \to \frac{1}{2}} E(\alpha, n_x, n_y, n_z) = \frac{\hbar c}{2} \frac{\pi}{2 \sin(\frac{\pi}{2a})} (n_x + n_y + n_z + 3) \tag{9.49}$$

In view of phenomenological shell models the fractional derivative parameter $\alpha$ allows a smooth analytic transition from a box- to a spherical symmetry.

If we interpret the degeneracy as a measure for the symmetry of a system, we obtain the result, that the symmetry of the fractional potential well is increasing for decreasing $\alpha$.

A typical area of application for the fractional potential well within the framework of nuclear physics is the use as a model potential for a shell model of nuclei.

The results, derived so far for the fractional potential well according to Fourier may be used in the case of the Riemann- and Caputo definition too.

The zeroes $k_n^0$ of the free solutions must be obtained in this case numerically. The eigenfunctions for the infinite potential well are then given by

$$\psi_{2n}^{(+)}(x) = {}_{\text{R,C}}\cos\left(\alpha, k_{2n}^0 \frac{x}{a}\right) \tag{9.50}$$

$$\psi_{2n+1}^{(-)}(x) = {}_{\text{R,C}}\sin\left(\alpha, k_{2n+1}^0 \frac{x}{a}\right) \tag{9.51}$$

And the corresponding eigenvalues are then given by:

$$e_n = \frac{1}{2}\left(\frac{\hbar}{mc}\right)^{2\alpha} mc^2 \left|\frac{k_n^0}{a}\right|^{2\alpha} \tag{9.52}$$

The extension to the multi-dimensional case reads:

$$\Psi_{n_1 n_2 \ldots n_N}(x^1, x^2, \ldots, x^N) = \prod_{i=1}^{N} \psi_{n_i}(x^i) \tag{9.53}$$

and the corresponding levels result as:

$$E_{n_1 n_2 \ldots n_N} = \frac{1}{2}\left(\frac{\hbar}{mc}\right)^{2\alpha} mc^2 \sum_{i=1}^{N} |\frac{k_{n_i}^0}{a_i}|^{2\alpha} \tag{9.54}$$

For $\alpha \approx 1$ the difference of the numerically determined energy spectrum and the analytic spectrum according to Fourier is small, as long as $\alpha \approx 1$.

When $\alpha$ becomes smaller, there is only a finite number of zeroes of the free fractional Schrödinger equation (see Fig. 7.1). As a consequence the energy spectrum is finite. For $\alpha < 0.65$ we obtain only one even and one odd solution for the Riemann-derivative, so we may occupy only the ground state and the first excited state of the spectrum. For the Caputo derivative we obtain only one solution and therefore only one energy level. Consequently, the number of particles, which may occupy the levels of the infinite potential well is limited. Of course, shell models used in nuclear physics, which intend to model a realistic behaviour, show a similar behaviour e.g. the Woods–Saxon-potential.

Hence we have presented exact solutions for the fractional Schrödinger equation with a potential term for the infinite well with Riemann- and Caputo derivative, which as a major result for $\alpha < 1$ generate only a finite number of energy levels. On the other hand the approximate solutions according to Fourier generate an infinite number of energy levels similar to the classical case.

It should be mentioned, that in the literature there are several presentations of so called exact solutions of the fractional Schrödinger equation with Coulomb potential [Laskin(2000)], potential wells [Laskin(2010)] and solved scattering problems [Guo(2006), Dong(2007)], which completely ignore the nonlocality of the fractional derivative and therefore should be considered merely as approximations valid for $\alpha \approx 1$ only [Jeng(2008)].

## 9.5 Radial solutions of the fractional Schrödinger equation

In the case of fractional derivative operators up to now no general theory of covariant coordinate transformations exists. Hence in the following we will only collect some arguments about aspects of a coordinate transform and make some remarks on radial solutions, which are independent of angular variables.

We intend to perform a coordinate transformation from Cartesian to hyperspherical coordinates in $\mathbb{R}^N$

$$f(x_1, x_2, ..., x_N) = f(r, \phi_1, \phi_2, ..., \phi_{N-1}) \tag{9.55}$$

The invariant line element in the case $\alpha = 1$

$$ds^2 = g_{ij} dx^i dx^j, \quad i, j = 1, ..., N \tag{9.56}$$

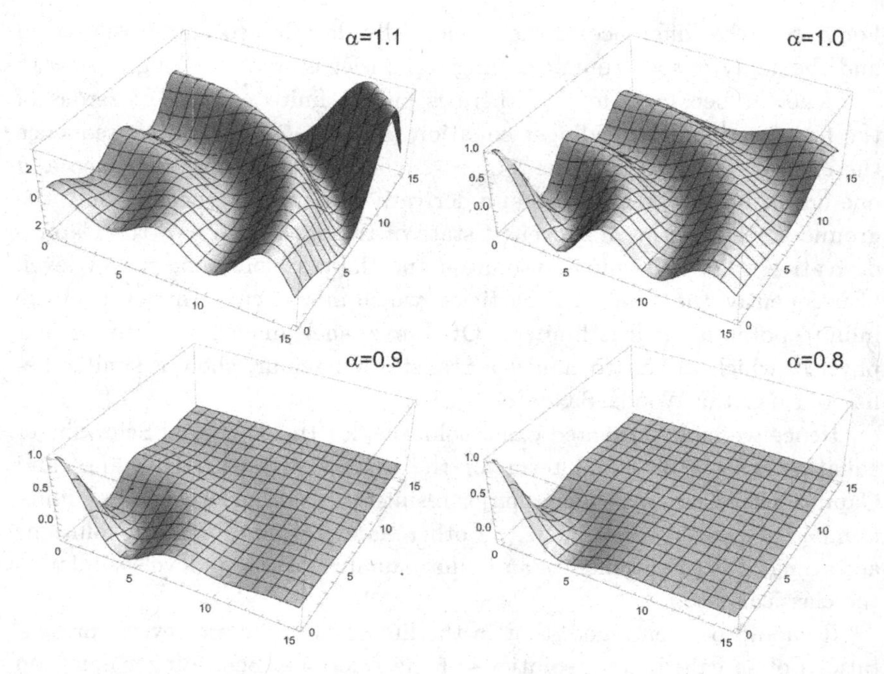


Fig. 9.5. Radial symmetric ground state solutions ($s = 0$) of the free fractional Schrödinger equation for different $\alpha$.

for arbitrary fractional derivative coefficient $\alpha$ may be generalized to

$$ds^{2\alpha} = g_{ij}^{\alpha} dx^{i\alpha} dx^{j\alpha}, \quad i,j = 1, ..., N \tag{9.57}$$

Consequently a reasonable definition of the radial coordinate is given by

$$r^{2\alpha} = \sum_{i=1}^{N} x_i^{2\alpha} \tag{9.58}$$

A radial symmetric solution is a function which only depends on the radial variable $r$. For fractional coordinates this does not imply, that this function is rotational symmetric.

But for a spherical ground state we require the following symmetry properties: We assume the spherical ground state to be independent of the angular variables, square integrable and of positive parity. If we apply the Caputo definition of the fractional derivative, therefore an appropriate ansatz is:

$$g(N, \alpha, kr) = \sum_{n=0}^{\infty} (-1)^n a_n(N, \alpha)(|k|r)^{2\alpha n} \tag{9.59}$$

or in Cartesian coordinates

$$g(N, \alpha, kx_1, ..., kx_N) = \sum_{n=0}^{\infty} (-1)^n a_n(N, \alpha) (\sum_{i=1}^{N} |kx|^{2\alpha})^n \tag{9.60}$$

where the coefficients $a_n$ depend on the explicit form of the potential. For a free particle, a solution on $\mathbb{R}^N$ is given with the abbreviation

$$\eta_j = \Gamma(1 + 2\alpha j)/\Gamma(1 + 2\alpha(j-1)) \tag{9.61}$$

by the recurrence relation

$$\begin{aligned} a_0 &= 1 \\ a_j &= a_{j-1}/\left((N-1)\, j\, \eta_1 + \eta_j\right), \qquad j = 1, 2, ... \end{aligned} \tag{9.62}$$

Following the naming conventions in atomic- and nuclear physics we call these solutions fractional s-waves. For the limit $\alpha = 1$ this solution reduces to a simple Bessel function.

Once these solutions are derived, the ground state solutions for a spherical potential well with infinite depth may be deduced accordingly:

An infinite spherical well is described by the potential

$$V(r) = \begin{cases} 0 & r \leq r_0 \\ \infty & r > r_0 \end{cases} \tag{9.63}$$

The corresponding boundary condition for the ground state wave function $g(N, \alpha, r)$ is:

$$g(N, \alpha, r_0) = 0 \tag{9.64}$$

Let $k^0_{\text{sph}}$ be the first zero of the free particle ground state wave function, the ground state wave function for the spherical infinite well potential is given by

$$g(N, \alpha, k^0_{\text{sph}} r/r_0) \tag{9.65}$$

and the ground state energy is then given by

$$e_0(N, \alpha) = \frac{1}{2} mc^2 \left( \frac{\hbar}{mc} \frac{k^0_{\text{sph}}}{r_0} \right)^{2\alpha} \tag{9.66}$$

If there are more than one zeroes, we may calculate the wave functions and energy values for higher (ns)-states.

From Fig. 9.5 we observe, that the radial symmetric solutions of the fractional Schrödinger equation and the behaviour of one dimensional solutions in Cartesian coordinates show a similar behaviour. In both cases the solutions show an increasing behaviour for $\alpha > 1$, while for $\alpha < 1$ they are located at the origin of the coordinate system.

We draw the conclusion, that coordinate transformations at the current state of the art are not an appropriate tool to investigate symmetry properties of fractional wave equations. As long as there is no general theory for fractional coordinate transformations, we should restrict investigations to Cartesian coordinate systems.

Since we are interested in fractional symmetry properties we apply a much more powerful tool: the fractional group theory. Indeed we will demonstrate in the following chapters the strength of a group theoretical approach and will use the concept of fractional group theory to investigate the properties of the fractional rotation group.

## Chapter 10

# Fractional Spin: a Property of Particles Described with the Fractional Schrödinger Equation

*Symmetries play an important role in physics. If a certain symmetry is present for a given problem, its solution will often be simplified [Greiner(1994), Frank(2009)].*

*For example homogeneity of space, which means, that space has equal structure at all positions $\vec{x}$ implies invariance under translations and therefore from a classical point of view conservation of momentum.*

*If we assume an isotropic space, which means, that space has a similar structure in all directions, this implies invariance of the system described under spatial rotations and consequently the conservation of angular momentum.*

*According to Noether's theorem, there always exists a conserved quantity, if the Euler–Lagrange equations are invariant under a specific coordinate transform [Noether(1918)].*

*In quantum mechanics a symmetry is described by an unitary operator $U$, which commutes with the Hamiltonian $H$. The existence of this symmetry operator implies the existence of a conserved observable. Let $G$ be the Hermitian generator of $U$:*

$$U = I - i\epsilon G + O(\epsilon^2) \tag{10.1}$$

*Since $U$ commutes with the Hamiltonian $H$, so does $G$:*

$$[H, G] = 0 \tag{10.2}$$

*and the conserved observable is the eigenvalue of $G$.*

*A fractional wave equation describes fractional particles, but the properties of space are not affected. We will therefore require, that symmetry properties of space will remain to be valid in the fractional case too. This implies, that e.g. the generator of a given symmetry operation should commute with the fractional Hamiltonian:*

$$[H^\alpha, G] = 0 \tag{10.3}$$

*As an example, we will investigate the properties of the fractional Schödinger equation under rotations and will unveil a quantity that is unexpected at first, that we will call fractional spin.*

## 10.1 Spin

Let us call a particle elementary, if it is described by a potential- and field-free wave equation. If in addition there is an internal structure, which is determined by additional quantum numbers, it may be revealed e.g. considering the behaviour under rotations.

The generator for infinitesimal rotations about the z-axis

$$L_z = i\hbar(x\partial_y - y\partial_x) \tag{10.4}$$

commutes with the Hamiltonian H of the free ordinary Schrödinger equation

$$[L_z, H] = 0 \tag{10.5}$$

which indicates, that the particle described has no internal structure.

But the same operator $L_z$ does not commute with the Dirac operator:

$$D = -i\hbar(\gamma^\mu \partial_\mu + m) \tag{10.6}$$

Instead it follows

$$[L_z, D] = \hbar^2(\gamma^x \partial_y - \gamma^y \partial_x) \tag{10.7}$$

Hence the invariance under rotations is lost. This is an indication for an internal structure of the particle described. There is an additional contribution, which results from the spin or intrinsic angular momentum of the particle.

In order to restore invariance of the wave equation under rotations we define the operator $J$ of total angular momentum with z-component $J_z$:

$$J_z = L_z + S_z \tag{10.8}$$

which indeed commutes with the Dirac-operator

$$[J_z, D] = 0 \tag{10.9}$$

For the z-component of the intrinsic angular momentum $S_z$ we obtain

$$S_z = -\frac{i\hbar}{4}[\gamma^x, \gamma^y] \tag{10.10}$$

and we conclude, that the Dirac equation may be used to describe the behaviour of spin-half particles.

## 10.2 Fractional spin

We will now proceed in a similar way to investigate the properties of particles, which are described with the fractional Schrödinger equation.

We will investigate the simplest case, the behaviour of the three dimensional free fractional Schrödinger equation

$$H^{\alpha}_{\text{free}}\psi = -\frac{1}{2}mc^2\left(\frac{\hbar}{mc}\right)^{2\alpha}\left(\partial_x^{2\alpha}+\partial_y^{2\alpha}+\partial_z^{2\alpha}\right)\psi = E\psi \tag{10.11}$$

under rotations in $R^2$ about the z-axis.

We start with a technical remark: According to the Leibniz product rule (3.39) the fractional derivative of the product $xf(x)$ is given by:

$$\begin{aligned}\partial_x^{\alpha}(xf(x)) &= \sum_{j=0}^{\infty}\binom{\alpha}{j}(\partial_x^j x)\partial_x^{\alpha-j}f(x) &(10.12)\\ &= \left(x\partial_x^{\alpha}+\alpha\partial_x^{\alpha-1}\right)f(x) &(10.13)\end{aligned}$$

We define a generalized fractional angular momentum operator $K^{\alpha}$ with z-component $K_z^{\alpha}$

$$K_z^{\alpha} = i\left(\frac{\hbar}{mc}\right)^{\alpha} mc\,(x\partial_y^{\alpha}-y\partial_x^{\alpha}) \tag{10.14}$$

The components $K_x^{\alpha}, K_y^{\alpha}$ are given by cyclic permutation of the spatial indices in (10.14). Using (10.12) the commutation relation with the Hamiltonian $H^{\alpha}_{\text{free}}$ of the free fractional Schrödinger equation (10.11) results as (setting $m=\hbar=c=1$):

$$\begin{aligned}[-i2K_z^{\beta}, H^{\alpha}_{\text{free}}] &= [K_z^{\beta}, \partial_x^{2\alpha}+\partial_y^{2\alpha}+\partial_z^{2\alpha}]\\ &= [K_z^{\beta}, \partial_x^{2\alpha}+\partial_y^{2\alpha}]\\ &= K_z^{\beta}(\partial_x^{2\alpha}+\partial_y^{2\alpha}) - (\partial_x^{2\alpha}+\partial_y^{2\alpha})K_z^{\beta}\\ &= (x\partial_y^{\beta}-y\partial_x^{\beta})(\partial_x^{2\alpha}+\partial_y^{2\alpha}) - (\partial_x^{2\alpha}+\partial_y^{2\alpha})(x\partial_y^{\beta}-y\partial_x^{\beta})\\ &= x\partial_x^{2\alpha}\partial_y^{\beta} + x\partial_y^{2\alpha+\beta} - y\partial_x^{2\alpha+\beta} - y\partial_x^{\beta}\partial_y^{2\alpha}\\ &\quad - \left(\partial_x^{2\alpha}x\partial_y^{\beta} - y\partial_x^{2\alpha+\beta} + x\partial_y^{2\alpha+\beta} - \partial_y^{2\alpha}y\partial_x^{\beta}\right)\\ &= x\partial_x^{2\alpha}\partial_y^{\beta} - y\partial_x^{\beta}\partial_y^{2\alpha}\\ &\quad - \left(x\partial_x^{2\alpha}\partial_y^{\beta} + 2\alpha\partial_x^{2\alpha-1}\partial_y^{\beta} - y\partial_x^{\beta}\partial_y^{2\alpha} - 2\alpha\partial_x^{\beta}\partial_y^{2\alpha-1}\right)\\ &= -2\alpha\left(\partial_x^{2\alpha-1}\partial_y^{\beta} - \partial_x^{\beta}\partial_y^{2\alpha-1}\right) \quad ,\alpha,\beta\in\mathbb{C} &(10.15)\end{aligned}$$

Setting $\beta=1$ we obtain for the z-component of the standard angular momentum operator $L_z$ (10.4) the commutation relation

$$[L_z, H^{\alpha}_{\text{free}}] = [K_z^{\beta=1}, H^{\alpha}_{\text{free}}] = -i\alpha\left(\partial_x^{2\alpha-1}\partial_y - \partial_x\partial_y^{2\alpha-1}\right) \tag{10.16}$$

which obviously is not vanishing. Therefore particles described with the fractional Schrödinger equation (9.19) carry an additional internal structure for $\alpha \neq 1$.

We now define the fractional total angular momentum $J^\beta$ with z-component $J_z^\beta$. Setting

$$\beta = 2\alpha - 1 \tag{10.17}$$

we obtain with $J_z^{2\alpha-1} = K_z^{2\alpha-1}$, and with (10.15)

$$[J_z^{2\alpha-1}, H_{\text{free}}^\alpha] = 0 \tag{10.18}$$

this operator commutes with the fractional free Hamiltonian. Therefore $J_z^{2\alpha-1}$ indeed is the fractional analogue of the standard z-component of the total angular momentum. Now we define the z-component of a fractional intrinsic angular momentum $S_x^\beta$ with

$$J_z^{2\alpha-1} = L_z + S_z^{2\alpha-1} \tag{10.19}$$

The explicit form is given by

$$S_z^{2\alpha-1} = i\left(x(\partial_y^{2\alpha-1} - \partial_y) - y(\partial_x^{2\alpha-1} - \partial_x)\right) \tag{10.20}$$

This operator vanishes for $\alpha = 1$, whereas for $\alpha \neq 1$ it yields the z-component of a fractional spin.

Lets call the difference between fractional and ordinary derivative $\delta p$, or more precisely the components

$$\delta p_i = i(\partial_i^{2\alpha-1} - \partial_i) \tag{10.21}$$

Hence for $S_z^{2\alpha-1}$ we can write

$$S_z^{2\alpha-1} = x\delta p_y - y\delta p_x \tag{10.22}$$

The components of a fractional spin vector are then given by the cross product

$$\vec{S}^{2\alpha-1} = \vec{r} \times \delta\vec{p} \tag{10.23}$$

Therefore fractional spin describes an internal fractional rotation, which is proportional to the momentum difference between fractional and ordinary momentum. For a given $\alpha$ it has exactly one component.

With $J_x^{2\alpha-1} = K_x^{2\alpha-1}$ and $J_y^{2\alpha-1} = K_y^{2\alpha-1}$, the commutation relations for the total fractional angular momentum are given by

$$\left[J_x^{2\alpha-1}, J_y^{2\alpha-1}\right] = (2\alpha - 1)\frac{\hbar}{mc}J_z^{2\alpha-1}p_z^{2(\alpha-1)} \tag{10.24}$$

$$\left[J_y^{2\alpha-1}, J_z^{2\alpha-1}\right] = (2\alpha - 1)\frac{\hbar}{mc}J_x^{2\alpha-1}p_x^{2(\alpha-1)} \tag{10.25}$$

$$\left[J_z^{2\alpha-1}, J_x^{2\alpha-1}\right] = (2\alpha - 1)\frac{\hbar}{mc}J_y^{2\alpha-1}p_y^{2(\alpha-1)} \tag{10.26}$$

with components of the fractional momentum operator $p^\alpha$ or generators of fractional translations respectively given as

$$p_i^\alpha = i \left( \frac{\hbar}{mc} \right)^\alpha mc \, \partial_i^\alpha \tag{10.27}$$

Therefore in the general case $\alpha \neq 1$ an extended fractional rotation group is generated, which contains an additional fractional translation factor.

We conclude, that fractional elementary particles which are described with the fractional Schrödinger equation, carry an internal structure, which we call fractional spin, because analogies to e.g. electron spin are close.

Consequently, the transformation properties of the fractional Schrödinger equation are more related to the ordinary Pauli equation than to the ordinary ($\alpha = 1$) Schrödinger equation.

# Chapter 11

# Factorization

*In the previous chapters we have presented a strategy to introduce a fractional extension for the non-relativistic Schrödinger equation. Within the framework of canonical quantization we replaced the classical observables* $\{x, p\}$ *by fractional derivative operators.*

*In this section we will present an alternative direct approach, which will automatically lead to fractional derivative wave equations.*

*Linearisation of a relativistic second order wave equation was first considered by Dirac [Dirac(1928)]. Starting with the relativistic Klein–Gordon equation the derived Dirac equation gave a correct description of the spin and the magnetic moment of the electron.*

*The concept of linearisation is important, since it provides a well defined mechanism to impose an additional SU(2) symmetry onto a given set of symmetry properties of a second order wave equation.*

*Since linearisation may be interpreted as a special case of factorization, namely to 2 factors, a natural generalization is a factorization to n factors.*

*Indeed we will find, that n-fold factorization of the Klein–Gordon equation automatically leads to fractional wave equations which describe particles with an internal* $SU(N)$*-symmetry.*

## 11.1 The Dirac equation

We start with the relativistic energy-momentum relation (setting $c = 1$)

$$E^2 = \vec{p}^2 + m^2 \tag{11.1}$$

Quantization with

$$E \to i\hbar\partial_t \tag{11.2}$$

$$p_i \to -i\hbar\partial_i \tag{11.3}$$

leads to the relativistic Klein–Gordon equation for particles with spin 0:

$$\left(-\hbar^2\partial^t\partial_t + \hbar^2\partial^i\partial_i - m^2)\right)\phi = 0 \tag{11.4}$$

$$\left(\hbar^2\partial^\mu\partial_\mu - m^2)\right)\phi = 0 \tag{11.5}$$

$$\left(\hbar^2\Box - m^2)\right)\phi = 0 \tag{11.6}$$

where we have introduced the d'Alembert-operator $\Box$.

Dirac proposed a two-fold factorization of the d'Alembert-operators introducing an operator $D$ with the property

$$D\,D' = \hbar^2\Box - m^2 \tag{11.7}$$

The explicit form of this operator is given by:

$$D = i\hbar\gamma^\mu\partial_\mu + m \tag{11.8}$$

where the $\gamma$ matrices fulfil the Clifford-algebra:

$$\gamma^\mu\gamma^\nu + \gamma^\nu\gamma^\mu = 2\delta^{\mu\nu} \tag{11.9}$$

where $\delta^{\mu\nu}$ is the Kronecker-delta.

The smallest representation of these $\gamma$-matrices is given as a triad of traceless $2\times 2$-Pauli-matrices $\sigma_i$:

$$\sigma_x = \begin{pmatrix} 0 & -i \\ i & 0 \end{pmatrix}, \quad \sigma_y = \begin{pmatrix} i & 0 \\ 0 & -i \end{pmatrix}, \quad \sigma_z = \begin{pmatrix} 1 & 0 \\ 0 & -1 \end{pmatrix} \tag{11.10}$$

These matrices are generators of the unitary group SU(2). The four $\gamma$-matrices are then generated using the external vector product

$$\gamma^0 = \mathbf{1}_2 \otimes \sigma_z, \qquad \gamma^i = \sigma_i \otimes \sigma_x, \quad i = 1, 2, 3 \tag{11.11}$$

where $\mathbf{1}_2$ is the $2\times 2$ unit matrix.

## 11.2 The fractional Dirac equation

We will now apply the method of factorization to the next complex case of three factors. Hence we search for an operator ${}^\alpha D$ with the property:

$${}^\alpha D\,{}^\alpha D'\,{}^\alpha D'' = \hbar^2\Box - m^2 \tag{11.12}$$

With the ansatz

$${}^\alpha D = \hbar^\alpha \gamma^\mu_\alpha \partial^\alpha_\mu + m^\alpha \tag{11.13}$$

where the index $\alpha$ indicates that the matrix dimensions depend on $\alpha$, the following conditions have to be fulfilled:

$$3\alpha = 2 \tag{11.14}$$

$$\gamma^{\mu}_{\alpha}\gamma^{\nu}_{\alpha}\gamma^{\sigma}_{\alpha}+\gamma^{\mu}_{\alpha}\gamma^{\sigma}_{\alpha}\gamma^{\mu}_{\alpha}+\gamma^{\nu}_{\alpha}\gamma^{\mu}_{\alpha}\gamma^{\sigma}_{\alpha}+\gamma^{\nu}_{\alpha}\gamma^{\sigma}_{\alpha}\gamma^{\mu}_{\alpha}+\gamma^{\sigma}_{\alpha}\gamma^{\mu}_{\alpha}\gamma^{\nu}_{\alpha}+\gamma^{\sigma}_{\alpha}\gamma^{\nu}_{\alpha}\gamma^{\mu}_{\alpha} = 6\,\delta^{\mu\nu\sigma} \tag{11.15}$$

And the explicit form of the triple factorized operator results as

$$^{2/3}D = \hbar^{2/3}\gamma^{\mu}_{2/3}\partial^{2/3}_{\mu} + m^{2/3}\mathbf{1}_9 \tag{11.16}$$

which is indeed a fractional wave equation. $\mathbf{1}_9$ is the $9 \times 9$ unit matrix, as we will demonstrate in the following.

The smallest representation of the $\gamma_{2/3}$-matrices is given in terms of traceless $3 \times 3$ matrices. We define the unit roots:

$$r_1 = e^{2\pi i \frac{1}{3}} = -\frac{1}{2} + \frac{\sqrt{3}}{2} i \tag{11.17}$$

$$r_2 = e^{2\pi i \frac{2}{3}} = -\frac{1}{2} - \frac{\sqrt{3}}{2} i \tag{11.18}$$

$$r_3 = e^{2\pi i} = 1 \tag{11.19}$$

A possible representation of the basis for all $3 \times 3$-matrices is then given by:

$$\begin{aligned}
\sigma_1 &= \begin{pmatrix} r_1 & 0 & 0 \\ 0 & r_2 & 0 \\ 0 & 0 & r_3 \end{pmatrix}, \quad \sigma_2 = \begin{pmatrix} r_2 & 0 & 0 \\ 0 & r_1 & 0 \\ 0 & 0 & r_3 \end{pmatrix}, \quad \sigma_3 = \begin{pmatrix} r_3 & 0 & 0 \\ 0 & r_3 & 0 \\ 0 & 0 & r_3 \end{pmatrix} \\
\sigma_4 &= \begin{pmatrix} 0 & r_1 & 0 \\ 0 & 0 & r_2 \\ r_3 & 0 & 0 \end{pmatrix}, \quad \sigma_5 = \begin{pmatrix} 0 & r_2 & 0 \\ 0 & 0 & r_1 \\ r_3 & 0 & 0 \end{pmatrix}, \quad \sigma_6 = \begin{pmatrix} 0 & r_3 & 0 \\ 0 & 0 & r_3 \\ r_3 & 0 & 0 \end{pmatrix} \\
\sigma_7 &= \begin{pmatrix} 0 & 0 & r_1 \\ r_2 & 0 & 0 \\ 0 & r_3 & 0 \end{pmatrix}, \quad \sigma_8 = \begin{pmatrix} 0 & 0 & r_2 \\ r_1 & 0 & 0 \\ 0 & r_3 & 0 \end{pmatrix}, \quad \sigma_9 = \begin{pmatrix} 0 & 0 & r_3 \\ r_3 & 0 & 0 \\ 0 & r_3 & 0 \end{pmatrix}
\end{aligned} \tag{11.20}$$

24 different triples of these matrices fulfil condition (11.15), namely:

$$\begin{gathered}
\{1,4,5\}\ \{1,4,6\}\ \{1,4,9\}\ \{1,5,6\}\ \{1,5,8\}\ \{1,6,7\}\ \{1,7,8\}\ \{1,7,9\} \\
\{1,8,9\}\ \{2,4,5\}\ \{2,4,6\}\ \{2,4,7\}\ \{2,5,6\}\ \{2,5,9\}\ \{2,6,8\}\ \{2,7,8\} \\
\{2,7,9\}\ \{2,8,9\}\ \{4,5,9\}\ \{4,6,7\}\ \{4,7,9\}\ \{5,6,8\}\ \{5,8,9\}\ \{6,7,8\}
\end{gathered} \tag{11.21}$$

6 of these 24 allowed combinations generate all three matrix diagonals, e.g. ($\{1,4,9\}$). Every single of these 24 triplets may be chosen as a realization for a set of $3 \times 3$ extended Pauli-matrices.

From an arbitrarily chosen set e.g. $\{1,4,5\}$ the follow extended Pauli matrices explicitly result as:

$$\sigma_x = \begin{pmatrix} 0 & r_1 & 0 \\ 0 & 0 & r_2 \\ r_3 & 0 & 0 \end{pmatrix}, \quad \sigma_y = \begin{pmatrix} 0 & r_2 & 0 \\ 0 & 0 & r_1 \\ r_3 & 0 & 0 \end{pmatrix}, \quad \sigma_z = \begin{pmatrix} r_1 & 0 & 0 \\ 0 & r_2 & 0 \\ 0 & 0 & r_3 \end{pmatrix} \tag{11.22}$$

and the four $\gamma_{2/3}$-matrices result as $9 \times 9$-matrices with the explicit form

$$\gamma^0_{2/3} = \mathbf{1}_3 \otimes \sigma_z \tag{11.23}$$

$$\gamma^i_{2/3} = \sigma_i \otimes \sigma_x, \qquad i = 1, 2, 3 \tag{11.24}$$

These matrices span a subspace of $SU(3)$.

Hence we have derived a fractional Dirac equation by a triple factorization of the d'Alembert operator. This procedure lead to two conditions: First a fractional derivative with $\alpha = 2/3$ is introduced. Second, the extended Clifford-algebra (11.15) determines the structure of the $\gamma$-matrices to be generators of $SU(3)$.

For the first time in the history of physics a formal method is proposed which derives a wave equation, which correctly describes the dynamics of a particle with internal $SU(n)$-symmetry.

The requirement of n-fold factorization of the d'Alembert-operator leads to fractional wave equations for $n > 2$.

To state this result more precisely: the extension of Diracs linearisation procedure which determines the correct coupling of a $SU(2)$ symmetric charge to a three-fold factorization of the d'Alembert-operator leads to a fractional wave equation with an inherent $SU(3)$ symmetry. This symmetry is formally deduced by the factorization procedure. In contrast to this formal derivation a standard Yang–Mills theory is merely a recipe, how to couple a phenomenologically deduced $SU(3)$ symmetry.

## 11.3 The fractional Pauli equation

There are two possible strategies to derive a non-relativistic limit of the fractional Dirac equation: We could try a fractional extension of a Foldy–Wouthuysen transformation or alternatively, we could factorize the standard non-relativistic Schrödinger equation.

Whether or not both methods lead to similar results, has not been examined yet.

In 1967, Levy–Leblond [Levy-Leblond(1967)] has linearised the standard non-relativistic Schrödinger equation and obtained a linear wave equation with an additional $SU(2)$ symmetry, but until now his approach has not been extended to higher order.

In a first approach we will derive the explicit form of a fractional operator $R$, which, iterated 3 times, conforms with the ordinary, non-relativistic

In the following we will show, that commutation relations may not be calculated independently of a specific function set any more. Instead, the results derived are only valid for a given function set, the operators act on.

As an illustrative example we will apply the fractional derivative according to Riemann and will calculate the explicit action of the commutation relations on the specific function set $|\nu\rangle = \hat{x}^{\alpha\nu}$. We obtain:

$$\begin{aligned} -i\hbar[\hat{x}^\alpha, {}_{\mathrm{R}}\hat{D}^\alpha]|\nu\rangle &= -i\hbar\left(\hat{x}^\alpha \frac{\Gamma(1+\nu\alpha)}{\Gamma(1+(\nu-1)\alpha)}|\nu-1\rangle - {}_{\mathrm{R}}\hat{D}^\alpha|\nu+1\rangle\right) \\ &= -i\hbar\left(\frac{\Gamma(1+\nu\alpha)}{\Gamma(1+(\nu-1)\alpha)} - \frac{\Gamma(1+(\nu+1)\alpha)}{\Gamma(1+\nu\alpha)}\right)|\nu\rangle \end{aligned} \tag{12.7}$$

With the abbreviation

$${}_{\mathrm{R}}c(\nu,\alpha) = \frac{\Gamma(1+(\nu+1)\alpha)}{\Gamma(1+\nu\alpha)} - \frac{\Gamma(1+\nu\alpha)}{\Gamma(1+(\nu-1)\alpha)} \tag{12.8}$$

the commutation relations (12.6) based on the fractional derivative according to Riemann result as:

$$[\hat{X}, {}_{\mathrm{R}}\hat{P}]\,|\nu\rangle = i\hbar_{\mathrm{R}}c(\nu,\alpha)\,|\nu\rangle \tag{12.9}$$

We obtain the important result that, in contrast to classical quantum mechanics, within the framework of fractional quantum mechanics all calculations depend on the specific representation of the Hilbert space, the fractional operators act on.

The minimum requirement for a fractional operator algebra is the explicit knowledge of the corresponding set of eigenfunctions.

Hence if we present the result of our derivation as

$$[\hat{X}, {}_{\mathrm{R}}\hat{P}] = i\hbar_{\mathrm{R}}c(\nu,\alpha) \tag{12.10}$$

this relation is valid only on the set of square integrable eigenfunctions $|\nu\rangle$.

For the same commutation relations with the Liouville definition of the fractional derivative on the set of eigenfunctions $|\nu\rangle = |x|^{-\alpha\nu}, x < 0$ we obtain:

$$\begin{aligned} -i\hbar[\hat{x}^\alpha, {}_{\mathrm{L}}\hat{D}^\alpha]|\nu\rangle &= -i\hbar\left(\hat{x}^\alpha \frac{\Gamma((\nu+1)\alpha)}{\Gamma(\nu\alpha)}|\nu+1\rangle - {}_{\mathrm{R}}\hat{D}^\alpha|\nu-1\rangle\right) \\ &= -i\hbar\,\mathrm{sign}(x)\left(|x|^\alpha \frac{\Gamma((\nu+1)\alpha)}{\Gamma(\nu\alpha)}|\nu+1\rangle - {}_{\mathrm{R}}\hat{D}^\alpha|\nu-1\rangle\right) \\ &= -i\hbar\,\mathrm{sign}(x)\left(\frac{\Gamma((\nu+1)\alpha)}{\Gamma(\nu\alpha)} - \frac{\Gamma(\nu\alpha)}{\Gamma((\nu-1)\alpha)}\right)|\nu\rangle \end{aligned} \tag{12.11}$$

and hence

$$ {}_{\mathrm{L}}c(\nu,\alpha) = \mathrm{sign}(x)\left(\frac{\Gamma(\nu\alpha)}{\Gamma((\nu-1)\alpha)} - \frac{\Gamma((\nu+1)\alpha)}{\Gamma(\nu\alpha)}\right) \quad x<0 \tag{12.12} $$

$$ = \frac{\Gamma((\nu+1)\alpha)}{\Gamma(\nu\alpha)} - \frac{\Gamma(\nu\alpha)}{\Gamma((\nu-1)\alpha)} \tag{12.13} $$

Finally, applying the fractional derivative according to Caputo the eigenfunctions of the commutation relations are given by $|\nu\rangle = x^{\nu\alpha}$ and therefore we may proceed

$$ {}_{\mathrm{C}}c(\nu,\alpha) = \begin{cases} \frac{\Gamma(1+(\nu+1)\alpha)}{\Gamma(1+\nu\alpha)} - \frac{\Gamma(1+\nu\alpha)}{\Gamma(1+(\nu-1)\alpha)} & \nu \neq 0 \\ \frac{\Gamma(1+(\nu+1)\alpha)}{\Gamma(1+\nu\alpha)} & \nu = 0 \end{cases} \tag{12.14} $$

Obviously the sets of eigenfunctions for the different commutation relations are eigenfunctions of the fractional one dimensional Euler operator $j_1^{\mathrm{E}} = \hat{x}^\alpha \hat{D}^\alpha$ as well. The corresponding eigenvalues follow as:

$$ {}_{\mathrm{R}}j_1^{\mathrm{E}} = \frac{\Gamma(1+\nu\alpha)}{\Gamma(1+(\nu-1)\alpha)} \tag{12.15} $$

$$ {}_{\mathrm{C}}j_1^{\mathrm{E}} = \begin{cases} \frac{\Gamma(1+\nu\alpha)}{\Gamma(1+(\nu-1)\alpha)} & \nu \neq 0 \\ 0 & \nu = 0 \end{cases} \tag{12.16} $$

$$ {}_{\mathrm{L}}j_1^{\mathrm{E}} = -\frac{\Gamma((\nu+1)\alpha)}{\Gamma(\nu\alpha)} \tag{12.17} $$

We conclude that an extension from standard group theory to fractional group theory is confronted with some additional difficulties. Despite this fact a group theoretical approach for a classification of multiplets of fractional extensions of standard groups will turn out to be a powerful tool to solve problems.

Especially non trivial symmetries in multi-dimensional spaces, like fractional rotational symmetry in $\mathbb{R}^3$ may successfully be investigated. Instead of solving the fractional Schödinger equation in hyperspherical coordinates directly, group theoretical methods provide elegant methods and strategies, to determine the eigenvalue spectrum of such highly symmetric problems analytically.

This will be explicitly shown in the next section.

## 12.2 The fractional rotation group $\mathrm{SO}_N^\alpha$

The components of the classical angular momentum in $\mathbb{R}^3$ are given by

$$ L_{ij} = x_i p_j - x_j p_i, \qquad i,j = 1,2,3 \tag{12.18} $$

The canonical quantization procedure (9.7), (9.8) determines the transition from classical mechanics to quantum mechanics we obtain the corresponding angular momentum operators

$$\hat{L}_{ij} = \hat{x}_i \hat{p}_j - \hat{x}_j \hat{p}_i, \qquad i,j = 1,2,3 \tag{12.19}$$

which are the generators of infinitesimal rotations in the $i, j$ plane.

Within the framework of fractional calculus we will propose a generalization to fractional rotations. For that purpose we replace in (12.19) according to (9.10) and (9.11) the classical operators by the fractional generalization

$$\hat{x}_i \to \hat{X}_i \tag{12.20}$$

$$\hat{p}_i \to \hat{P}_i \tag{12.21}$$

and define the generators of infinitesimal fractional rotations in the $i, j$ plane in $\mathbb{R}^N$ with $(i, j = 1, ..., 3N)$ as:

$$\begin{aligned} L^\alpha_{ij} &= \hat{X}_i \hat{P}_j - \hat{X}_j \hat{P}_i \\ &= -i\hbar \left( \hat{x}^\alpha_i \hat{D}^\alpha_j - \hat{x}^\alpha_j \hat{D}^\alpha_i \right) \end{aligned} \tag{12.22}$$

In order to calculate the commutation relations of the fractional angular momentum operators, we introduce the function set $f$:

$$\{ f(\hat{x}^\alpha_1, \hat{x}^\alpha_2, ..., \hat{x}^\alpha_{3N}) = \prod_i^{3N} \hat{x}^{\nu\alpha}_i \} \tag{12.23}$$

On this function set the commutation relations for the generators $L^\alpha_{ij}$ are isomorphic to the fractional extension $SO^\alpha(3N)$ of the standard rotation group $SO(3N)$:

$$[L^\alpha_{ij}, L^\alpha_{mn}] = i\hbar\, c(\nu, \alpha) \left( \delta_{im} L^\alpha_{jn} + \delta_{jn} L^\alpha_{im} - \delta_{in} L^\alpha_{jm} - \delta_{jm} L^\alpha_{in} \right) \tag{12.24}$$

where $c(\nu, \alpha)$ denotes the commutation relation

$$c(\nu, \alpha) = [\hat{x}^\alpha_i, \hat{D}^\alpha_i], \qquad i = 1, ..., 3N \tag{12.25}$$

Hence we may derive the eigenvalue spectrum of the fractional angular momentum operators in a similar manner as for the standard rotation group [Louck(1972)].

We define a set of Casimir operators $\Lambda^2_k$, where the index $k$ indicates the Casimir operator associated with $SO^\alpha(k)$:

$$\Lambda^2_k = \frac{1}{2} \sum_{i,j}^{k} (L^\alpha_{ij})^2, \qquad k = 2, ..., 3N \tag{12.26}$$

Indeed these operators fulfil the commutation relations:

$$[\Lambda^2_{3N}, L^\alpha_{ij}] = 0 \tag{12.27}$$

and successively

$$[\Lambda^2_k, \Lambda^2_{k'}] = 0. \tag{12.28}$$

The multiplets of $SO^\alpha(3N)$ may be therefore classified according to the group chain

$$SO^\alpha(3N) \supset SO^\alpha(3N-1) \supset \ldots \supset SO^\alpha(3) \supset SO^\alpha(2) \tag{12.29}$$

We use Einstein's summation convention and introduce the metric of the Euclidean space to be $g_{ij} = \delta_{ij}$ for raising and lowering indices.

The explicit form of the Casimir operators is then given by

$$\Lambda^2_k = +\hat{X}^i\hat{X}_i\hat{P}^j\hat{P}_j - i\hbar\, c(\nu,\alpha)\,(k-1)\hat{X}^j\hat{P}_j - \hat{X}^i\hat{X}^j\hat{P}_i\hat{P}_j \qquad i,j = 1,\ldots,k \tag{12.30}$$

The classical homogeneous Euler operator is defined as $x_1\partial_{x_1} + x_2\partial_{x_2} + \ldots + x_k\partial_{x_k}$.

We introduce a generalisation of the classical homogeneous Euler operator $J^e_k$ for fractional derivative operators

$$J^e_k = \hat{x}^{\alpha\, i}\, \hat{D}^\alpha_i, \qquad i = 1,\ldots,k \tag{12.31}$$

With the generalised homogeneous Euler operator the Casimir operators are given explicitly:

$$\Lambda^2_k = +\hat{X}^i\hat{X}_i\hat{P}^j\hat{P}_j + \hbar^2\Big(c(\nu,\alpha)\,(k-2)J^e_k + J^e_kJ^e_k\Big), \quad i,j = 1,\ldots,k \tag{12.32}$$

From this equation follows, that the Casimir operator is diagonal on a function set $f$, if the generalised homogeneous Euler operator is diagonal on $f$ and if $f$ fulfils the Laplace-equation

$$\hat{D}^{\alpha\, i}\hat{D}^\alpha_i f = 0 \tag{12.33}$$

We will show, that the generalised homogeneous Euler operator is diagonal, if $f$ fulfils an extended fractional homogeneity condition, which we will derive in the following. For that purpose, we first verify, that the generalised homogeneous Euler operator is diagonal on a homogeneous function in the one dimensional case.

As an example we will apply the Riemann fractional derivative. For other derivative definitions the procedure is similar.

With $\hat{\lambda}^\alpha = \text{sign}(\lambda)\,|\lambda|^\alpha$ the following scaling property, which is valid for the fractional derivative [Oldham(1974)] will be used:

$$\hat{D}_{\hat{\lambda}^\alpha} f(\hat{\lambda}^\alpha \hat{x}^\alpha) = \hat{x}^\alpha \hat{D}^\alpha_{\hat{\lambda}^\alpha \hat{x}^\alpha} f(\hat{\lambda}^\alpha \hat{x}^\alpha) \qquad (12.34)$$

Homogeneity of a function in one dimension implies:

$$f(\hat{\lambda}^\alpha \hat{x}^\alpha) = \hat{\lambda}^{\alpha\,\nu} f(\hat{x}^\alpha), \qquad \nu\alpha > -1 \qquad (12.35)$$

We apply the Riemann derivative to (12.35) with (12.34) we obtain:

$$\hat{x}^\alpha \hat{D}^\alpha_{\hat{\lambda}^\alpha \hat{x}^\alpha} f(\hat{\lambda}^\alpha \hat{x}^\alpha) = \frac{\Gamma(1+\nu\alpha)}{\Gamma(1+(\nu-1)\alpha)} \hat{\lambda}^{\alpha(\nu-1)} f(\hat{x}^\alpha) \qquad (12.36)$$

Which for $\hat{\lambda}^\alpha = 1$ reduces (12.36) to:

$$\hat{x}^\alpha \hat{D}^\alpha_{\hat{x}^\alpha} f(\hat{x}^\alpha) = \frac{\Gamma(1+\nu\alpha)}{\Gamma(1+(\nu-1)\alpha)} f(\hat{x}^\alpha) \qquad (12.37)$$

The left hand side of (12.37) is nothing else but the generalised homogeneous Euler operator (12.31) in one dimension $J^e_{k=1}$.

Hence we have demonstrated that a one dimensional homogeneous function is an eigenfunction of the fractional Euler operator.

Therefore, for the multi-dimensional case, on a function set of the form

$$\{f(\hat{x}^\alpha_1, \hat{x}^\alpha_2, ..., \hat{x}^\alpha_k) = \sum_{\nu_i} a_{\nu_1 \nu_2 ... \nu_k} \hat{x}_1^{\alpha\,\nu_1} \hat{x}^{\alpha\,\nu_2} ... \hat{x}_k^{\alpha\,\nu_k}\} \qquad (12.38)$$

the Euler operator $J^e_k$ is diagonal, if the $\nu_i$ fulfil the following condition:

$$\sum_{i=1}^{k} \frac{\Gamma(1+\nu_i\alpha)}{\Gamma(1+(\nu_i-1)\alpha)} = \frac{\Gamma(1+\nu\alpha)}{\Gamma(1+(\nu-1)\alpha)} \qquad (12.39)$$

This is the fractional homogeneity condition.

Hence we define the Hilbert space $H^\alpha$ of all functions $f$, which fulfil the fractional homogeneity condition (12.39), satisfy the Laplace equation $\bar{D}^i \bar{D}_i f = 0$ and are normalised in the interval $[-1, 1]$.

We propose the quantization condition:

$$\nu = n, \qquad n = 0, 1, 2, ... \qquad (12.40)$$

Where $n$ is a non-negative integer. This specific choice reduces to the classical quantisation condition for the case $\alpha = 1$.

On this Hilbert space $H^\alpha$, the generalised homogeneous Euler operator $J^e_k$ is diagonal and has the eigenvalues $l_k(\alpha, n)$

$$l_k(\alpha, n) = \frac{\Gamma(1+n\alpha)}{\Gamma(1+(n-1)\alpha)}, \qquad n = 0, 1, 2, ... \qquad (12.41)$$

The eigenvalues of the Casimir operators on $H^\alpha$ follow as:

$$\Lambda_2 f = \hbar l_2(\alpha, n) f \tag{12.42}$$

$$\Lambda_k^2 f = \hbar^2 l_k(\alpha, n)\Big(l_k(\alpha, n) + c(n, \alpha)\,(k-2)\Big) f, \quad k = 3, ..., 3N \tag{12.43}$$

with

$$l_k(\alpha, n) \geq l_{k-1}(\alpha, n) \geq ... \geq \mid \pm l_2(\alpha, n) \mid \geq 0 \tag{12.44}$$

We have derived an analytic expression for the full spectrum of the Casimir operators for the $SO^\alpha(3N)$, which may be interpreted as a projection of function set (12.38) on to function set (12.23) determined by the fractional homogeneity condition (12.39).

For the special case of only one particle ($N = 1$), we can introduce the quantum numbers $J$ and $M$, which now denote the $J$-th and $M$-th eigenvalue $l_3(\alpha, J)$ and $l_2(\alpha, M)$ of the generalised homogeneous Euler operators $J_3^e$ and $J_2^e$ respectively.

The eigenfunctions are fully determined by these two quantum numbers $f = \mid JM\rangle$.

With the definitions $\hat{J}_z(\alpha) = \Lambda_2 = L_{12}^\alpha$ and $\hat{J}^2(\alpha) = \Lambda_3^2 = L_{12}^{\alpha\,2} + L_{13}^{\alpha\,2} + L_{23}^{\alpha\,2}$ it follows

$$\hat{J}_z(\alpha) \mid JM\rangle = \pm\hbar \frac{\Gamma(1 + |M|\alpha)}{\Gamma(1 + (|M| - 1)\alpha)} \mid JM\rangle \qquad M = \pm 0, \pm 1, \pm 2, ..., \pm J \tag{12.45}$$

$$\hat{J}^2(\alpha) \mid JM\rangle = \hbar^2 \frac{\Gamma(1 + (J+1)\alpha)}{\Gamma(1 + (J-1)\alpha)} \mid JM\rangle \qquad J = 0, +1, +2, ... \tag{12.46}$$

For $\alpha = 1$ equations (12.45) and (12.46) reduce to the classical values

$$\hat{J}_z(\alpha = 1) \mid JM\rangle = \hbar M \mid JM\rangle, \qquad M = 0, \pm 1, \pm 2, ..., \pm J \tag{12.47}$$

$$\hat{J}^2(\alpha = 1) \mid JM\rangle = \hbar^2 J(J+1) \mid JM\rangle, \qquad J = 0, +1, +2, ... \tag{12.48}$$

and $\hat{J}^2(\alpha = 1)$, $\hat{J}_z(\alpha = 1)$ reduce to the classical operators $\hat{J}^2$, $\hat{J}_z$ used in standard quantum mechanical angular momentum algebra [Edmonds(1957), Rose(1995)].

The complete eigenvalue spectrum of the Casimir operators of the fractional rotation group has been calculated algebraically for the Riemann fractional derivative.

It is important to mention, that there are two different eigenvalues for $M = 0$ as a consequence of the non vanishing derivative of a homogeneous function of degree $n = 0$. This is a special property of the Riemann fractional derivative definition.

To illustrate this point we present the Cartesian representation of the eigenfunction $f$ for the simple case $\alpha = \frac{1}{2}$ and $n = 2$ explicitly:

$$f(\alpha = 1/2, n = 2, x, y) = \mid \pm 0\rangle = y^{-\frac{1}{2}} \pm i x^{-\frac{1}{2}} \tag{12.49}$$

Indeed it follows:

$$l_2(\alpha = 1/2, n = 2) \mid \pm 0\rangle = \frac{1}{\Gamma(1/2)} \mid \pm 0\rangle \tag{12.50}$$

$$L_z^{\alpha=1/2} \mid \pm 0\rangle = \pm \frac{1}{\Gamma(1/2)} \mid \pm 0\rangle \tag{12.51}$$

The calculation of the eigenvalue spectrum of the Casimir operators for different fractional derivative definitions follows the same procedure.

On the basis of the Caputo derivative we obtain for the eigenvalue spectrum of the fractional Euler operator:

$$l_k(\alpha, n) = \begin{cases} \frac{\Gamma(1+n\alpha)}{\Gamma(1+(n-1)\alpha)} & n = 1, 2, ... \\ 0 & n = 0 \end{cases} \tag{12.52}$$

which differs from the Riemann definition only in the case $n = 0$ and for the sequence of eigenvalues for the Casimir operators we obtain:

$$\hat{J}_z(\alpha) \mid JM\rangle = \begin{cases} \pm\hbar \frac{\Gamma(1+|M|\alpha)}{\Gamma(1+(|M|-1)\alpha)} \mid JM\rangle & M = \pm 1, \pm 2, ..., \pm J \\ 0 & M = 0 \end{cases} \tag{12.53}$$

$$\hat{J}^2(\alpha) \mid JM\rangle = \begin{cases} \hbar^2 \frac{\Gamma(1+(J+1)\alpha)}{\Gamma(1+(J-1)\alpha)} \mid JM\rangle & J = +1, +2, ... \\ 0 & J = 0 \end{cases} \tag{12.54}$$

Therefore we have classified the multiplets of the fractional extension of the standard rotation group $SO^\alpha(3N)$.

Now we may search for a realisation of $SO^\alpha(3)$ symmetry in nature. Promising candidates for a still unrevealed $SO^\alpha(3)$ symmetry are the ground state band spectra of even-even nuclei. Therefore we will introduce the fractional symmetric rigid rotor model in the next chapter and discuss some of its applications.

# Chapter 13

# The Fractional Symmetric Rigid Rotor

*As a first application of the fractional rotation group the ground state excitation spectra of even-even nuclei will be successfully interpreted with a fully analytic fractional symmetric rigid rotor model.*

*In a generalised, unique approach the fractional symmetric rigid rotor treats rotations, the $\gamma$-unstable limit, vibrations and the linear potential limit similarly as fractional rotations are all included in the same symmetry group, the fractional $SO^\alpha(3)$. This is an encouraging unifying point of view and a new powerful approach for the interpretation of nuclear ground state band spectra.*

*The results derived may be applied to other branches of physics as well, e.g. molecular spectroscopy.*

## 13.1 The spectrum of the fractional symmetric rigid rotor

The fractional stationary Schrödinger equation is given as

$$H\Psi = E\Psi \tag{13.1}$$

The requirement of invariance under fractional rotations in $\mathbb{R}^3$ completely determines the structure of the fractional Hamiltonian up to a constant $A_0$:

$$H = m_0 + A_0 \Lambda_3^2 \left(\mathrm{SO}^\alpha(3)\right) \tag{13.2}$$

$$= m_0 + A_0 \hat{J}^2(\alpha) \tag{13.3}$$

where the constant $m_0$ acts as a counter term for the zero-point contribution of the fractional rotational energy and constant $A_0$ is a measure for the level spacing.

In the last section we have demonstrated, that the Hamiltonian is diagonal on a function set $|JM\rangle$ diagonal.

Furthermore the Hamiltonian commutes with the parity operator $\Pi$, $[H, \Pi] = 0$.

For $\alpha = 1$ the function set $|JM\rangle$ reduces to the set of spherical harmonics $Y_{JM}$ in Cartesian representation. The eigenvalues of the parity operator $\Pi$ are then given by:

$$\begin{aligned} \Pi Y_{JM}(x_1, x_2, x_3) &= Y_{JM}(-x_1, -x_2, -x_3) \\ &= (-1)^J Y_{JM}(x_1, x_2, x_3) \end{aligned} \tag{13.4}$$

Hence the wave function $Y_{JM}$ is invariant under parity transformation $\Pi$, if $J$ is restricted to even, non negative integers $J = 0, 2, 4, ...$. In a collective geometric model, this symmetry is interpreted as the geometry of the symmetric rigid rotor model [Eisenberg(1987)].

Whether or not the behaviour (13.4) is still valid for the function set $|JM\rangle$ with arbitrary $\alpha$ cannot be proven directly with the methods developed so far.

Nevertheless, restricting $J$ to be an even, non-negative integer $J = 0, 2, 4, ...$ for arbitrary $\alpha$, implies a symmetry, which we call in analogy to the case $\alpha = 1$ the fractional symmetric rigid rotor, even though this term lacks a direct geometric interpretation for $\alpha \neq 1$.

Hence we define the spectrum of a fractional symmetric rigid rotor:

$$E = E_J^\alpha = m_0 + A_0 \hbar^2 \frac{\Gamma\Big(1 + (J+1)\alpha\Big)}{\Gamma\Big(1 + (J-1)\alpha\Big)}, \qquad J = 0, 2, 4, ... \tag{13.5}$$

where we have used the Riemann version of the fractional Casimir operator. We will present convincing arguments, why this variant of a fractional derivative is the right choice for the fractional rigid rotor spectrum. In Fig. 13.1 this level spectrum is plotted as a function of $\alpha$.

As a general trend the higher angular momentum energy values are decreasing for $\alpha < 1$. This behaviour is a well established observation for nuclear low energy rotational band structures [Eisenberg(1987)].

In a classical geometric picture of the nucleus [Bohr(1952)], [Eisenberg(1987)] this phenomenon is interpreted as a change of the nuclear shape under rotations, causing an increasing moment of inertia and therefore a decrease of rotational bands.

Within the framework of the fractional model presented so far the moment of inertia is fixed (since $A_0$ is a constant) but the same effect now

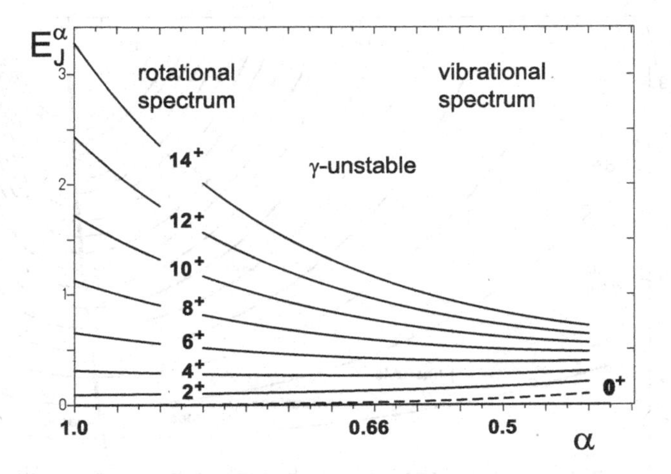


Fig. 13.1. The energy level spectrum $E_J^\alpha$ according to (13.5) for the fractional symmetric rigid rotor. The dotted line $0^+$ shows the ground state energy for the Riemann fractional derivative. If we had used the Caputo derivative instead, the ground state energy was to be exactly 0.

results as an inherent property of the fractional derivatives angular momentum.

Another remarkable characteristic of the fractional symmetric rigid rotor spectrum is due to the fact, that the relative spacing between different levels is changing with $\alpha$ (see Fig. 13.2).

In classical geometric models there are distinct analytically solvable limits, e.g. the rotational, vibrational and so called $\gamma$-unstable limits.

These limits are characterised by their independence of the potential energy from $\gamma$. In that case, the five dimensional Bohr Hamiltonian is separable [Bohr(1952), Fortunato(2005)].

With the collective coordinates $\beta$, $\gamma$ and the three Euler angles $\theta_i$ the product ansatz for the wave function

$$\Psi(\beta,\gamma,\theta_i) = f(\beta)\Phi(\gamma,\theta_i) \tag{13.6}$$

leads to a differential equation for $\beta$, expressed in canonical form:

$$\left\{\frac{\hbar^2}{2B}\Big(-\frac{\partial^2}{\partial\beta^2} + \frac{(\tau+1)(\tau+2)}{\beta^2}\Big) + V(\beta)\right\}\Big(\beta^2 f(\beta)\Big) \tag{13.7}$$
$$= E(n_\beta,\tau)\Big(\beta^2 f(\beta)\Big)$$

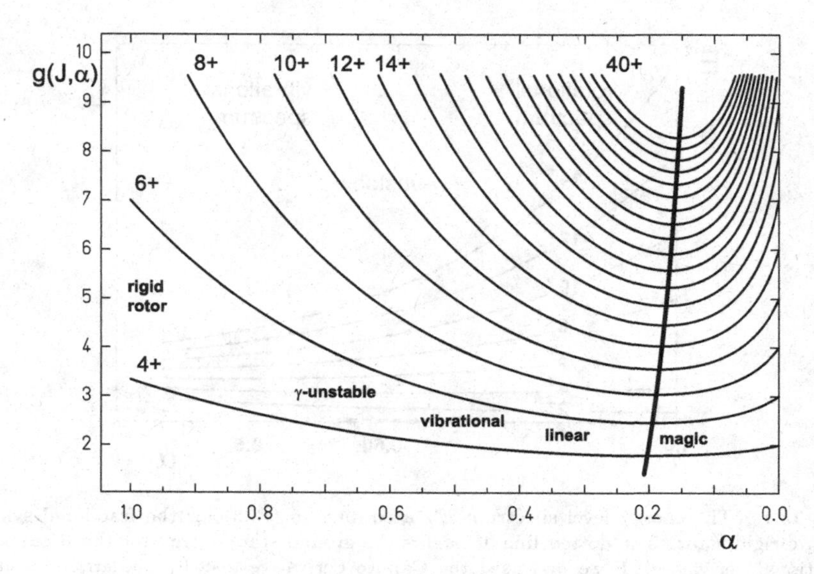


Fig. 13.2. Relative energy levels $g(J,\alpha) = (E_J^\alpha - E_0^\alpha)/(E_2^\alpha - E_0^\alpha)$ with $E_J^\alpha$ from (13.5) for the fractional symmetric rigid rotor. The thick line indicates the minimum ratio for a given $J$ as a function of $\alpha$.

The ground state spectra of nuclei, based on $\gamma$ independent potentials, are then determined by the conditions [Fortunato(2005)]:

$$n_\beta = 0 \tag{13.8}$$
$$2\tau = J, \qquad \tau = 0, 1, 2, ...$$

In the following we will prove, that the above mentioned limits are included within the fractional symmetric rigid rotor spectrum as special cases at distinct $\alpha$-values. This is a further indication, that the fractional symmetric rigid rotor model may be successfully applied to low energy excitation spectra of nuclei.

## 13.2 Rotational limit

In the geometric collective model the rotational limit is described by the symmetric rigid rotor [Bohr(1954), Eisenberg(1987)]:

$$E = \frac{1}{2}\frac{\hbar^2}{\Theta} J(J+1), \qquad J = 0, 2, 4, ... \tag{13.9}$$

This limit is trivially included in the fractional symmetric rigid rotor spectrum for $\alpha = 1$.

$$E_J^{\alpha=1} = m_0 + A_0\hbar^2 \frac{\Gamma\Big(1+(J+1)\Big)}{\Gamma\Big(1+(J-1)\Big)} \tag{13.10}$$

$$= m_0 + A_0\hbar^2 \frac{(J+1)!}{(J-1)!}$$

$$= m_0 + A_0\hbar^2 J(J+1), \qquad J = 0, 2, 4, ...$$

Setting $m_0 = 0$ and $A_0 = 1/(2\Theta)$ completes the derivation.

Remarkably enough, reading (13.9) backwards is an application of the original Euler concept for a fractional derivative, since Euler suggested a replacement of $n!$ for the discrete case by $\Gamma(1+n)$ for the fractional case [Euler(1738)].

## 13.3 Vibrational limit

In the geometric collective model the vibrational limit is described by a harmonic oscillator potential [Eisenberg(1987)]

$$V(\beta) = \frac{1}{2}C\beta^2 \tag{13.11}$$

where $C$ is the stiffness in $\beta$-direction. The level spectrum is given by

$$E = \hbar\omega(N + 5/2) \tag{13.12}$$

with $\omega = \sqrt{C/B}$ and

$$N = 2n_\beta + \tau, \qquad N = 0, 1, 2, 3, ... \tag{13.13}$$

Therefore the ground state band $E^{g.s.}$ is given according to the conditions (13.8) by

$$E^{g.s.}(\tau) = \hbar\omega(\tau + 5/2), \qquad \tau = 0, 1, 2, ... \tag{13.14}$$

We will now prove, that for $\alpha = 1/2$ the spectrum for the fractional symmetric rigid rotor corresponds to this vibrational type ground state spectrum

$$E_J^{\alpha=1/2} = m_0 + A_0\hbar^2 \frac{\Gamma\Big(1+\frac{1}{2}(J+1)\Big)}{\Gamma\Big(1+\frac{1}{2}(J-1)\Big)} \tag{13.15}$$

$$= m_0 + A_0\hbar^2 \frac{\Gamma(1+J/2+\frac{1}{2})}{\Gamma(1+J/2-\frac{1}{2})}$$

$$= m_0 + A_0\hbar^2 \frac{\Gamma(1+J/2+\frac{1}{2})}{\Gamma(J/2+\frac{1}{2})}$$

$$= m_0 + A_0\hbar^2 \left(J/2 + \frac{1}{2}\right), \qquad J = 0, 2, 4, ...$$

where we have used $\Gamma(1+z) = z\Gamma(z)$.

Since $J = 2\tau$ for the ground state band we get:

$$E_J^{\alpha=1/2} = m_0 + A_0\hbar^2 \left(\tau + \frac{1}{2}\right), \qquad \tau = 0, 1, 2, ... \tag{13.16}$$

Setting $m_0 = 2\hbar\omega$ and $A_0 = \omega/\hbar$ completes the derivation.

It should at least be mentioned, that the above few lines mark the realization of an old alchemist's dream: the transmutation of a given group to another (here from $SO(n)$ to $U(n)$). This was also the ambitious aim of q-deformed Lie algebras [Gupta(1992)], but was missed until now.

The appearance of the correct equidistant level spacing including the zero-point energy contribution of the harmonic oscillator eigenvalues (13.16) is due to the fact, that the Riemann fractional derivative does not vanish when applied to a constant function. Instead, for the fractional symmetric rigid rotor based on the Riemann fractional derivative there always exists a zero-point energy for $\alpha \neq 1$ of the form

$$\hat{J}^2(\alpha)\,|00\rangle = \hbar^2 \frac{\Gamma(1+\alpha)}{\Gamma(1-\alpha)}\,|00\rangle \tag{13.17}$$

If we had chosen the Caputo derivative definition, for $\alpha = 1/2$ we would have obtained an equidistant spectrum too, but the zero-point energy would have been $E_0^{\alpha=1/2} = 0$ and consequently would have been too small.

Hence the consistence with the spectrum of the harmonic oscillator including the zero-point energy is a strong argument for our specific choice of the Riemann fractional derivative definition.

## 13.4 Davidson potential: the so called $\gamma$-unstable limit

A third analytically solvable case exists, which is based on the Davidson potential [Davidson(1932)], which originally was proposed to describe an interaction in diatomic molecules.

The potential is of the form

$$V(\beta) = \frac{1}{8} C\,\beta_0^2 \left(\frac{\beta}{\beta_0} - \frac{\beta_0}{\beta}\right)^2 \tag{13.18}$$

where $\beta_0$ (the position of the minimum) and $C$ (the stiffness in $\beta$-direction at the minimum) are the parameters of the model.

For $\beta_0 = 0$ this potential is equivalent to the harmonic oscillator potential.

In the general case, the energy level spectrum is given by

$$E(n_\beta, \tau) = \hbar\omega \left( n_\beta + \frac{1}{2} + \frac{1}{2} a_\tau \right) - \frac{1}{4} C \beta_0^2 \tag{13.19}$$

with

$$a_\tau = \frac{1}{2}\sqrt{\sqrt{BC}\beta_0^4 + (2\tau + 3)^2} \tag{13.20}$$

For the ground state band we get

$$E^{g.s.}(\tau) = \hbar\omega \left( \frac{1}{2} + \frac{1}{2} a_\tau \right) - \frac{1}{4} C \beta_0^2 \tag{13.21}$$

For $\beta_0 > 0$ we expand the square root in $a_\tau$ in a Taylor-series

$$a_\tau = \frac{1}{2}\sqrt{\sqrt{BC}\beta_0^2}\,(1 + \frac{1}{2}\frac{(2\tau + 3)^2}{\sqrt{BC}\beta_0^4} + ...) \tag{13.22}$$

Shifting $\tau$ by

$$\tau = \hat{\tau} - 3/2 \tag{13.23}$$

causes the linear term to vanish and the resulting level scheme is of the form

$$E^{g.s.}(\hat{\tau}) = c_0 + c_1 \hat{\tau}^2 + ... \tag{13.24}$$

Therefore the $\gamma$-unstable Davidson potential is characterised by the condition, that the linear term in a shifted series expansion in $\tau$ vanishes.

In order to determine the corresponding fractional coefficient $\alpha$, we shift the fractional energy spectrum by $-3/2$ and expand in a Taylor series at $J/2 = \tau = 0$ ($\Psi$ and $\Psi^1$ denote the di- and tri-gamma function):

$$\begin{aligned} E^\alpha_{\tau - \frac{3}{2}} &= m_0 + A_0 \hbar^2 \frac{\Gamma(1 - \alpha/2)}{\Gamma(1 - 5\alpha/2)} \\ &\times \Bigg[ 1 - \alpha \Big( \Psi(1 - 5\alpha/2) - \Psi(1 - \alpha/2) \Big) \tau \\ &+ \frac{\alpha^2}{2} \Bigg\{ \Big( \Psi(1 - 5\alpha/2) - \Psi(1 - \alpha/2) \Big)^2 \\ &- \Big( \Psi^1(1 - 5\alpha/2) - \Psi^1(1 - \alpha/2) \Big) \Bigg\} \tau^2 + ... \Bigg] \end{aligned} \tag{13.25}$$

The linear term in $\tau$ has to vanish. Therefore $\alpha$ is determined by the condition $\alpha$:

$$\Psi(1 - 5\alpha/2) = \Psi(1 - \alpha/2) \tag{13.26}$$

which is fulfilled for $\alpha \approx 0.66$.

Hence for a sufficiently large $\beta_0$ the ground state band spectrum of the $\gamma$-unstable Davidson potential is reproduced within the fractional symmetric rigid rotor at $\alpha \approx 2/3$.

## 13.5 Linear potential limit

The hitherto discussed three special cases, which are known to be analytically solvable in geometric and IBM-models respectively were found to be realized within proposed fractional rotor model using a fractional derivative parameter in the range $1/2 \le \alpha \le 1$. But the domain of allowed fractional derivative parameters is at least extendible down to $\alpha = 0$. Therefore our interest is attracted to fractional derivative parameters with $\alpha < 1/2$. As an intuitive guess we may assume, that this region may be interpreted in terms of a geometric model using potentials which increase weaker than a harmonic oscillator potential:

$$V(\beta) \sim |\beta|^{\nu} \qquad \nu < 2, \nu \in \mathbb{R} \tag{13.27}$$

In view of fractional calculus therefore the investigation of the properties of e.g. a linear potential is an obvious extension of the well established geometric model potentials. Since geometric models are subject of interest for more than 50 years now, it is surprising that potentials of the type (13.27) did not play any role and were never considered even for academic reasons only.

An investigation of potentials of type (13.27) is a natural consequence of a fractional extension of $\mathrm{SO}^{\alpha}(N)$. For geometric models this subject is an open task. This is a nice example for the fact, that on one hand even in this research area there are still open questions and on the other hand, that a concept based on fractional calculus indeed may stimulate research in other areas.

Hence the linear potential was proposed and studied theoretically as late as 2005 by Fortunato [Fortunato(2005)].

Although it has not been used for a description of nuclear ground state band spectra yet, it will turn out to be useful for an understanding of nuclear spectra near closed shells discussed in the following section.

The potential is of the form

$$V(\beta) = C|\beta| \tag{13.28}$$

where $C$ (the stiffness in $\beta$-direction) is the main parameter of the model.

The level spectrum is given approximately by [Fortunato(2005)]:

$$E(n_\beta, \tau) = \frac{3}{2}C\beta_0 + (n_\beta + \frac{1}{2})\sqrt{\frac{3C}{\beta_0}} \tag{13.29}$$

with

$$\beta_0 = \Big(2(\tau + 1)(\tau + 2)/C\Big)^{1/3} \tag{13.30}$$

The ground state band level spectrum is given according to conditions (13.8) by

$$E^{g.s.}(\tau) = \frac{3}{2}C\beta_0 + \frac{1}{2}\sqrt{\frac{3C}{\beta_0}} \tag{13.31}$$

We therefore are able to define the relative energy levels $f(\tau)$ in units $E^{g.s.}(1) - E^{g.s.}(0)$, which are independent of parameter C

$$f(\tau) = \frac{E^{g.s.}(\tau) - E^{g.s.}(0)}{E^{g.s.}(1) - E^{g.s.}(0)} \tag{13.32}$$

An expansion in a Taylor series at $\tau = 1$ yields:

$$f(\tau) = 1 + 0.895(\tau - 1) - 0.076(\tau - 1)^2 + 0.018(\tau - 1)^3 + \ldots \tag{13.33}$$

An equivalent expression for the fractional rotational energy is given by:

$$g(\tau, \alpha) = \frac{E^\alpha_{2\tau} - E^\alpha_0}{E^\alpha_2 - E^\alpha_0}, \qquad \tau \in \mathbb{N} \tag{13.34}$$

A Taylor series expansion at $\tau = 1$ followed by a comparison of the linear terms of $f(\tau)$ and $g(\tau, \alpha)$ leads to the condition

$$\frac{2\alpha\Gamma(1-\alpha)\Gamma(3\alpha)\Big(2 + 3\alpha\Psi(\alpha) - \Psi(3\alpha)\Big)}{\Gamma^2(1+\alpha) - \Gamma(1-\alpha)\Gamma(1+3\alpha)} = 0.895 \tag{13.35}$$

There are two solutions given by:

$$\alpha_1 = 0.33 \tag{13.36}$$

$$\alpha_2 = 0.11 \tag{13.37}$$

Hence below the vibrational region at $\alpha \approx 1/2$ there exists a region at $\alpha \approx 1/3$, which corresponds to the linear potential model in a geometric picture.

## 13.6 The magic limit

In Fig. 13.2 an additional region is visible, which attracts our attention. Near $\alpha \approx 0.2$ a minimum occurs for the relative level spacing. Of course this immediately raises the question, if we can assign a model potential in this case too.

The answer will be given in terms of an approximation of the effective potential in terms of the shifted harmonic oscillator:

Starting with a model potential of the form

$$V(\beta) = C\beta^\nu, \qquad \nu \geq 0 \tag{13.38}$$

according to (13.7) the effective potential is given by:

$$V_{\text{eff}}(\beta) = C\beta^{\nu} + \frac{(\tau+1)(\tau+2)}{\beta^2} \tag{13.39}$$

The minimum of the potential at $\beta_0$ may be calculated from the requirement:

$$\frac{d}{d\beta} V_{\text{eff}}(\beta) \,|_{\beta=\beta_0} = 0 \tag{13.40}$$

We obtain

$$\beta_0 = \left( \frac{2(\tau+1)(\tau+2)}{\nu C} \right)^{\frac{1}{\nu+2}} \tag{13.41}$$

A series expansion of the effective potential at $\beta_0$ up to second order in $\beta$ leads to:

$$V_{\text{eff}}(\beta) \approx c_0 + \frac{1}{2}\omega^2(\beta - \beta_0)^2 \tag{13.42}$$

The constants $c_0$ and $\omega^2$ result as

$$c_0 = C\left(1 + \frac{\nu}{2}\right)\beta_0^{\nu} \tag{13.43}$$

$$\omega^2 = C\,\frac{\nu(\nu+2)}{2}\,\beta_0^{\nu-2} \tag{13.44}$$

Energy eigenvalues in units $\hbar$ are then given as

$$E(n_\beta, \tau, n) = c_0 + \left(n_\beta + \frac{1}{2}\right)\omega \tag{13.45}$$

In Fig. 13.3 we present the single terms of the effective potential and the harmonic approximation for different angular momenta $\tau$.

For the ground state band we obtain according to conditions (13.8)

$$E^{g.s.}(\tau) = c_0 + \frac{1}{2}\omega \tag{13.46}$$

The special case of the linear potential ($\nu = 1$) presented in the previous section has been investigated by Fortunato using the same method of harmonic approximation.

Let us present the relative energy values $f(\tau)$ in units $E^{g.s.}(1) - E^{g.s.}(0)$, which are independent of the parameter $C$.

$$f(\tau, \nu) = \frac{E^{g.s.}(\tau) - E^{g.s.}(0)}{E^{g.s.}(1) - E^{g.s.}(0)} \tag{13.47}$$

The minimum for the relative level spacing for the fractional rigid rotor can be deduced from Fig. 13.2 as $\alpha \approx 0.2$ and $R(4^+/2^+) = 1.79$ results.

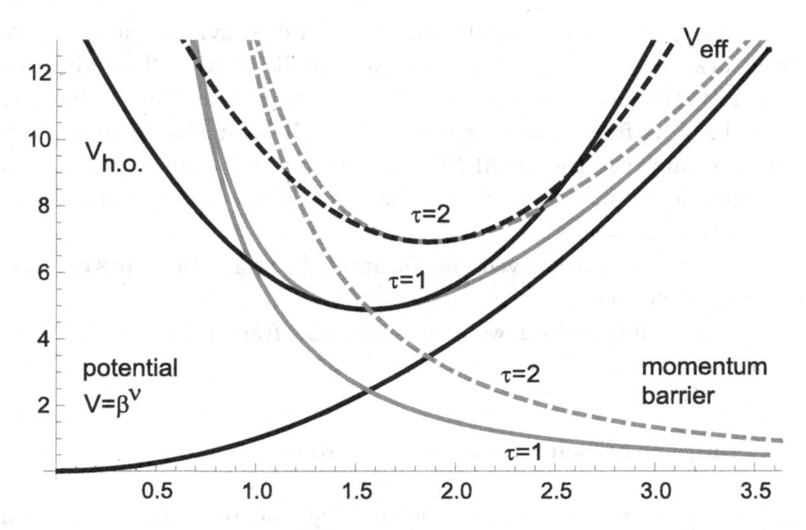


Fig. 13.3. An approximation for the geometric model potential $V = \beta^\nu$ for an analytic calculation of the level spectrum. Gray lines are the angular momentum barrier, the sum of these two terms which is the effective potential $V_{\text{eff}}$ and the the harmonic approximation $V_{\text{h.o.}}$. Thick lines correspond to $\tau = 1$ and dashed lines indicate $\tau = 2$.

Within the framework of the geometric model we now will search for a value for $\nu$ which leads to a similar ratio.

Indeed it follows that in the limit $\nu \to 0$ the ratio

$$\lim_{\nu \to 0} f(\tau = 2^+, \nu) = 1.72 \tag{13.48}$$

We obtain the remarkable result, that we may associate within the limitations of the analytic harmonic approximation the minimum of the relative level spacing at $\alpha \approx 0.2$ with a vanishing collective potential in the geometric model.

In other words, if the macroscopic collective potential vanishes, the spectrum is dominated by microscopic effects like shell- and pairing corrections. This situation is indeed realized near atomic nuclei with magic proton and neutron numbers respectively.

Therefore the validity of fractional symmetric rigid rotor model by far exceeds the scope of a collective geometric model.

Let us summarize the results:

A change of the fractional derivative coefficient $\alpha$ may be interpreted within a geometric collective model as a change of the potential energy surface.

In a generalised, unique approach the fractional symmetric rigid rotor treats rotations at $\alpha \approx 1$, the $\gamma$-unstable limit at $\alpha \approx 2/3$, vibrations at $\alpha \approx 1/2$, the linear potential limit at $\alpha \approx 1/3$ and even the magic limit similarly as fractional rotations. They all are included in the same symmetry group, the fractional $SO^\alpha(3)$. This is an encouraging unifying point of view and a new powerful approach for the interpretation of nuclear ground state band spectra.

Of course the results derived may be applied to other branches of physics as well, e.g. molecular spectroscopy.

In the following section we will apply the fractional rotor model to experimental data.

## 13.7 Comparison with experimental data

In the previous section we have shown, that the rotational, vibrational and $\gamma$-unstable limit of geometric collective models are special cases of the fractional symmetric rigid rotor spectrum.

The fractional derivative coefficient $\alpha$ acts like an order parameter and allows a smooth transition between these idealised cases.

Of course, a smooth transition between rotational and e.g. vibrational spectra may be achieved by geometric collective models too. A typical example is the Gneuss–Greiner model [Gneuss(1971)] with a more sophisticated potential. Critical phase transitions from vibrational to rotational states have been studied for decades using e.g. coherent states formalism or within the framework of the IBM-model [Casten(1993), Feng(1981), Haapakoski(1970), Raduta(2005, 2010)]. But in general, within these models, results may be obtained only numerically with extensive effort, while the fractional rotor model leads to analytic results for intermediate cases too.

Especially the intermediate cases are important, because there are only very few nuclei, whose spectra may correspond to the idealized limits.

In the following we will prove, that the full range of low energy ground state band spectra of even-even nuclei is accurately reproduced within the framework of the fractional symmetric rigid rotor model.

For a fit of the experimental spectra $E_J^{\text{exp}}$ we will use (13.5)

$$E_J^\alpha = m_0 + a_0 \frac{\Gamma\Big(1+(J+1)\alpha\Big)}{\Gamma\Big(1+(J-1)\alpha\Big)}, \qquad J = 0, 2, 4, \ldots \tag{13.49}$$

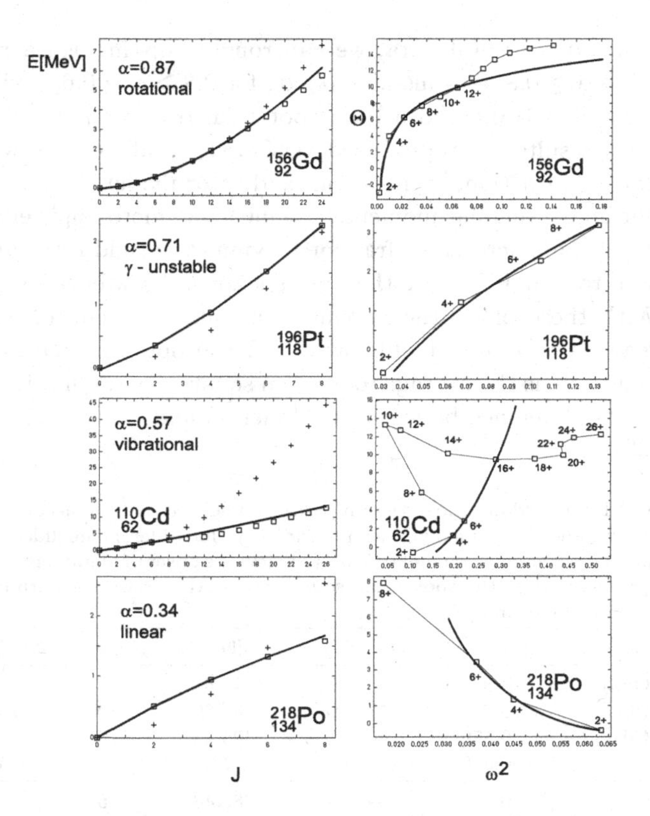


Fig. 13.4. The left column shows energy levels of the ground state bands for $^{156}$Gd, $^{196}$Pt, $^{110}$Cd and $^{218}$Po for increasing $J$. Squares indicate the experimental values, +-symbols indicate the optimum fit of the classical symmetric rigid rotor and the curves give the best fit for the fractional symmetric rigid rotor. In the right column the corresponding back bending plots are shown.

with the slight modification, that $\hbar^2$ has been included into the definition of $a_0$, so that $m_0$ and $a_0$ may both be given in units of [keV].

As a first application we will analyse typical rotational, $\gamma$-unstable, vibrational and linear type spectra.

In the upper row of Fig. 13.4 the energy levels of the ground state bands are plotted for $^{156}$Gd, $^{196}$Pt, $^{110}$Cd and $^{218}$Po which represent typical rotational-, $\gamma$-unstable-, vibrational and linear type spectra.

The fractional coefficients $\alpha$, deduced from the experimental data, are remarkably close to the theoretically expected idealised limits of the fractional symmetric rigid rotor.

From the experimental data, we can roughly distinguish a rotational region for $1 \geq \alpha \geq 0.8$, a $\gamma$-unstable region for $0.8 \geq \alpha \geq 0.6$, a vibrational region for $0.6 \geq \alpha \geq 0.4$ and a linear potential region for $0.4 \geq \alpha \geq 0.2$. Experimental results are reproduced very well for all regions within the framework of the fractional symmetric rigid rotor model.

Since for higher angular momenta commencing microscopic effects limit the validity of the macroscopic fractional symmetric rigid rotor model fits, in the lower row of Fig. 13.4 the corresponding back-bending plots are shown. With these plots, the maximum angular momentum $J_{max}$ for a valid fit may be determined. In Table 13.1 the optimum parameter sets $(\alpha, a_0, m_0)$ as well as the average root mean square deviation $\Delta E$ are listed. In general, the difference between experimental and calculated energies is less than 2%.

Table 13.1. Listed are the optimum parameter sets ($\alpha$, $a_0$, $m_0$ according to (13.49)) for the fractional symmetric rigid rotor for different nuclides. The maximum valid angular momentum $J_{max}$ below the onset of alignment effects are given as well as the root mean square error $\Delta E$ between experimental and fitted energies in %.

| nuclid | $\alpha$ | $a_0$[keV] | $m_0$[keV] | $J_{max}$ | $\Delta E$[%] |
|---|---|---|---|---|---|
| ${}^{156}_{92}Gd_{64}$ | 0.863 | 31.90 | −43.65 | 14 | 2.23 |
| ${}^{196}_{118}Pt_{78}$ | 0.710 | 175.06 | −83.69 | 10 | 0.44 |
| ${}^{110}_{62}Cd_{48}$ | 0.570 | 607.91 | −405.80 | 6 | 0 |
| ${}^{218}_{134}Po_{84}$ | 0.345 | 1035.69 | −671.03 | 8 | 0.12 |
| ${}^{164}_{88}Os_{76}$ | 0.624 | 339.423 | −128.448 | 6 | 0 |
| ${}^{170}_{94}Os_{76}$ | 0.743 | 125.128 | −39.968 | 10 | 1.35 |
| ${}^{172}_{96}Os_{76}$ | 0.767 | 89.010 | −23.656 | 8 | 1.21 |
| ${}^{174}_{98}Os_{76}$ | 0.771 | 63.388 | −17.656 | 24 | 0.73 |
| ${}^{176}_{100}Os_{76}$ | 0.808 | 51.231 | −23.064 | 18 | 1.22 |
| ${}^{178}_{102}Os_{76}$ | 0.816 | 49.882 | −22.792 | 14 | 2.16 |
| ${}^{180}_{104}Os_{76}$ | 0.841 | 45.309 | −18.662 | 12 | 2.50 |
| ${}^{182}_{106}Os_{76}$ | 0.903 | 32.002 | −7.159 | 10 | 1.29 |
| ${}^{184}_{108}Os_{76}$ | 0.904 | 31.690 | −12.902 | 14 | 1.39 |
| ${}^{186}_{110}Os_{76}$ | 0.882 | 40.433 | −19.155 | 14 | 1.82 |
| ${}^{188}_{112}Os_{76}$ | 0.875 | 44.567 | −14.629 | 12 | 1.58 |
| ${}^{190}_{114}Os_{76}$ | 0.847 | 58.055 | −17.748 | 12 | 1.34 |
| ${}^{192}_{116}Os_{76}$ | 0.835 | 63.774 | −15.437 | 12 | 0.71 |
| ${}^{214}_{126}Ra_{88}$ | −0.007 | 374408 | −376529 | 8 | 2.53 |
| ${}^{214}_{126}Ra_{88}$ | 0.548 | 344.107 | 305.766 | 24 | 8.27 |
| ${}^{216}_{128}Ra_{88}$ | 0.181 | 3665.36 | −2887.88 | 10 | 4.22 |
| ${}^{218}_{130}Ra_{88}$ | 0.536 | 321.622 | −160.787 | 30 | 1.25 |
| ${}^{220}_{132}Ra_{88}$ | 0.696 | 83.603 | −25.966 | 30 | 0.26 |
| ${}^{222}_{134}Ra_{88}$ | 0.831 | 33.221 | −5.67 | 6 | 0 |
| ${}^{224}_{136}Ra_{88}$ | 0.841 | 27.295 | −8.849 | 12 | 1.57 |

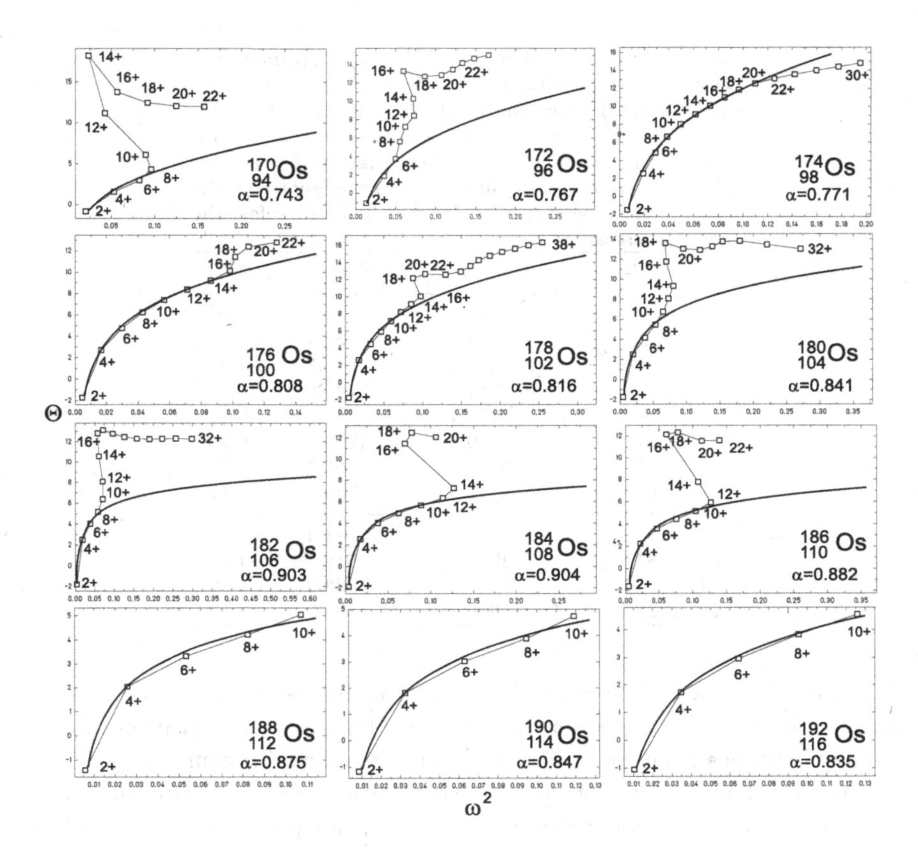


Fig. 13.5. For a set of osmium isotopes back-bending plots are shown. Squares indicate experimental values. The thick line is the optimum fit result for the fractional symmetric rigid rotor. Optimum fit parameter sets are given in Table 13.1. For the fit all angular momenta below the onset of microscopic alignment effects are included.

This means, that the error is better than a factor 3 to 6 compared to the results of a standard Taylor series expansion up to second order in $J$.

As a second application of the fractional symmetric rigid rotor, we study systematic isotopic effects for osmium isotopes. Optimum fit parameter sets are given in the second part of Table 13.1, the corresponding back-bending plots are given in Fig. 13.5. In Table 13.2 the results for the fitted ground state band energies of the fractional symmetric rigid rotor model according to (13.49) are compared to the experimental levels for $^{176}$Os.

Obviously below the beginning of alignment effects the spectra are reproduced very well within the fractional symmetric rigid rotor model. The

Table 13.2. Energy levels for $^{176}$Os with optimum parameter set from Table 13.1 for the fractional symmetric rigid rotor $E_J^\alpha$ according to (13.49) compared with the experimental values $E_J^{\text{exp}}$ and relative error $\Delta$E in % for different angular momenta $J$. Note that for $J > 14\hbar$ beginning microscopic alignment effects are ignored in the fractional symmetric rigid rotor model and therefore the error is increasing.

| $J[\hbar]$ | $E_J^\alpha[keV]$ | $E_J^{\text{exp}}[keV]$ | $\Delta$E[%] |
|---|---|---|---|
| $0^+$ | −13.1 | 0.0 | — |
| $2^+$ | 144.9 | 135.1 | −7.29 |
| $4^+$ | 404.4 | 395.3 | −2.31 |
| $6^+$ | 744.8 | 742.3 | 0.33 |
| $8^+$ | 1155.4 | 1157.5 | 0.17 |
| $10^+$ | 1629.6 | 1633.8 | 0.25 |
| $12^+$ | 2162.3 | 2167.7 | 0.24 |
| $14^+$ | 2749.8 | 2754.6 | 0.17 |
| $(16^+)$ | 3389.2 | 3381.4 | −0.23 |
| $(18^+)$ | 4077.8 | 4019.1 | −1.46 |
| $(20^+)$ | 4813.6 | 4683.2 | −2.78 |
| $(22^+)$ | 5594.8 | 5398.8 | −3.63 |

fractional coefficient $\alpha$ listed in Table 13.1 increases slowly for increasing nucleon numbers. Consequently, we observe a smooth transition of the osmium isotopes from the $\gamma$-unstable to the rotational region.

As a third application, we investigate the systematics of the ground state energy level structure near closed shells. For that purpose, in Fig. 13.6 a series of spectra of radium isotopes near the magic neutron number $N = 126$ are plotted.

Starting with radium $^{224}$Ra, the full variety of possible spectral types emerges while approaching the magic $^{214}$Ra.

While $^{224}$Ra and $^{222}$Ra are pure rotors, $^{220}$Ra shows a perfect $\gamma$-unstable type spectrum. $^{218}$Ra presents a spectrum of a almost ideal vibrational type. $^{216}$Ra is even closer to the magic $^{214}$Ra and represents a new class of nuclear spectra.

In a geometric picture, this spectrum may be best interpreted as a linear potential spectrum as proposed in Sec. 13.5. Typical candidates for this kind of spectrum are nuclei in close vicinity of magic numbers like $^{218}_{134}\text{Po}_{84}$, $^{154}_{84}\text{Yb}_{70}$, $^{134}_{80}\text{Xe}_{54}$, $^{96}_{56}\text{Zr}_{40}$ or $^{88}_{50}\text{Sr}_{38}$.

Finally, the experimental ground state spectrum of the magic nucleus $^{214}$Ra shows a clustering of energy values: for low angular momenta $\alpha$ tends towards zero, which becomes manifest through almost degenerated

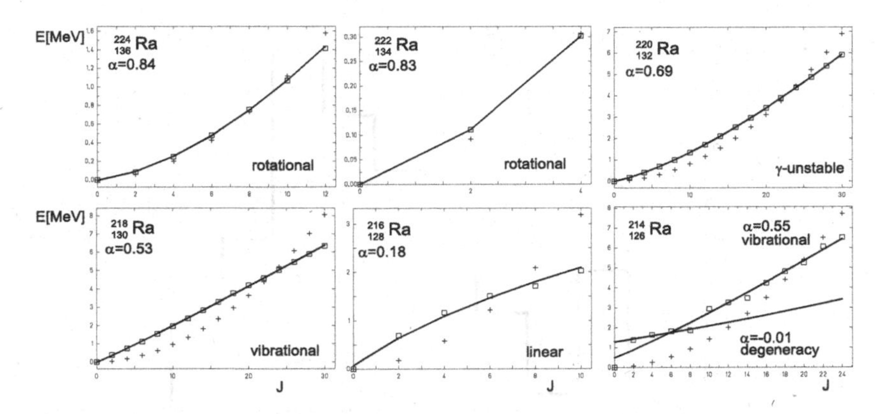


Fig. 13.6. The change of energy level structure of ground state bands near closed shells is illustrated for the magic neutron number $N = 126$ with a set of radium isotopes. Squares indicate experimental energy values. Crosses indicate the optimum fit with the standard symmetric rigid rotor model. The thick line is the optimum fit result for the fractional symmetric rigid rotor. Optimum fit parameter sets are given in the lower part of Table 13.1. For $^{214}$Ra, two different fits for low and higher angular momentum indicate the different spectral regions for magic nucleon numbers.

energy levels while for higher angular momenta the spectrum tends to the vibrational type. In a classical picture this is interpreted as the dominating influence of microscopic shell effects. Within the framework of catastrophe theory this observation could be interpreted as a bifurcation as well.

Therefore the fractional symmetric rigid rotor model is not well suited for a description of the full spectrum of a nucleus with magic proton or neutron numbers. But all other nuclear spectra are described with a high grade of accuracy within the framework of the fractional symmetric rigid rotor.

As a remarkable fact $\alpha$ reduces smoothly within the interval $1 \geq \alpha \geq 0$ while approaching a magic number, e.g. $N = 126$. Therefore, $\alpha$ is an appropriate order parameter for such sequences of ground state band spectra.

As a fourth application we analyse some general aspects, which are common to all even-even nuclei. Experimental ground state band spectra up to at least $J = 4^+$ are currently known for 490 even-even nuclei. For these nuclei a full parameter fit according (13.49) was performed. The resulting distribution $n(\alpha)$ of fitted $\alpha$ values in 0.1 steps is plotted in Fig. 13.7.

This distribution is not evenly spread. Most of even-even nuclei exhibit rotational type ground state band spectra followed by $\gamma$-unstable and

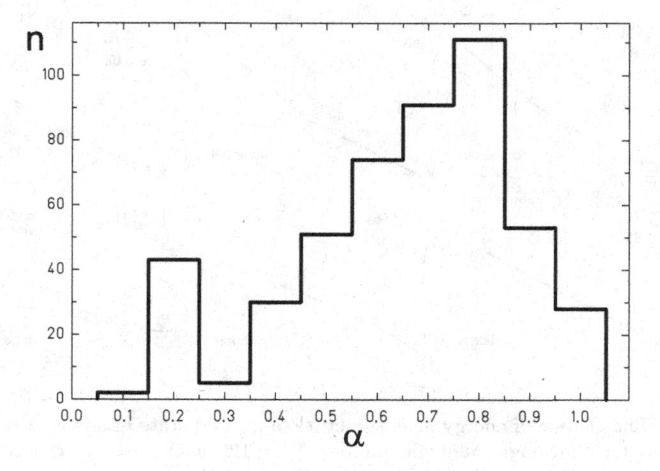


Fig. 13.7. The distribution $n(\alpha)$ of $\alpha$-values from a fit of 490 ground state band spectra of even-even nuclei, accumulated in intervals $\Delta\alpha = 0.1$ from ${}^{8}_{4}\mathrm{Be}_{4}$ to ${}^{256}_{156}\mathrm{Fm}_{100}$ and ${}^{254}_{152}\mathrm{No}_{102}$ respectively.

vibrational type spectra. There is a distribution gap at $\alpha \approx 0.3$. This is the region, where in a geometrical picture nuclear spectra corresponding to the linear potential model are expected.

About 8% of even-even nuclear spectra are best described with $\alpha \approx 0.2$. The sequence of radium isotopes, presented in Fig. 13.6 reflects the general distribution of spectra for even-even nuclei.

An interesting feature may be deduced from the observation, that $n(\alpha)$ is almost linear in the region $0.3 \leq \alpha \leq 0.8$.

Hence, with the proton number $Z$ and the neutron number $N$, we define the distance $R(Z, N)$ in the $(Z, N)$-plane, from the next magic proton or neutron number respectively to be

$$R(Z, N) = \sqrt{(Z - Z_{\text{magic}})^2 + (N - N_{\text{magic}})^2} \tag{13.50}$$

The linearity of $n(\alpha)$ implies a linear dependence of $\alpha$ from $R$

$$\alpha(Z, N) = c_0 + c_1 R(Z, N), \qquad 0.25 \leq \alpha \leq 0.85 \tag{13.51}$$

which is a helpful relation to determine an estimate of $\alpha$ for a series of nuclei.

Therefore the series of $\alpha$ values for osmium isotopes, given in Table 13.1 may be understood even quantitatively:

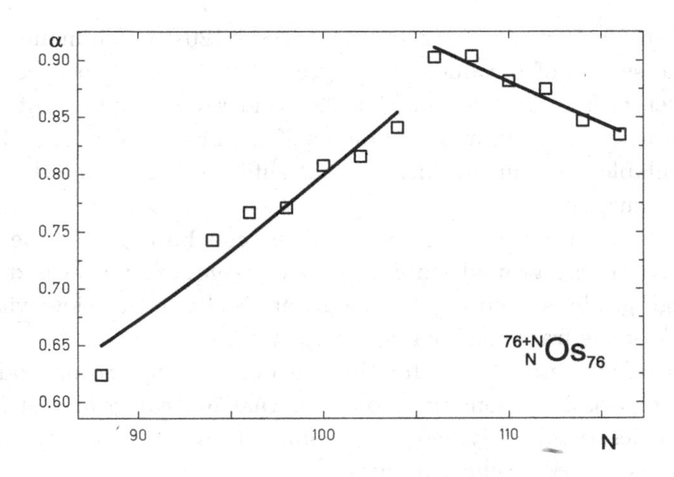


Fig. 13.8. Plot of fitted $\alpha$ parameters for a set of osmium isotopes as a function of increasing neutron number $N$. Squares indicate the optimum $\alpha$ listed in Table 13.1. Thick lines are fits according to equation (13.51).

The series of ${}^{164}_{88}$Os to ${}^{182}_{104}$Os is closer to the $N = 82$ magic shell and therefore shows an increasing sequence in $\alpha$. A least-squares fit yields $c_0 = 0.528$ and $c_1 = 0.0143$ with an average error in $\alpha$ less than 3%

The series ${}^{184}_{106}$Os to ${}^{192}_{116}$Os is closer to the $N = 126$ shell and therefore shows a descending sequence in $\alpha$. A least-squares fit yields $c_0 = 0.744$ and $c_1 = 0.008$ with an error in $\alpha$ less than 1%.

In Fig. 13.8 these results are presented graphically. From the figure we can deduce a prediction for $\alpha({}^{166}_{90}\text{Os}) = 0.671$ and $\alpha({}^{168}_{92}\text{Os}) = 0.695$. From these predicted $\alpha$ values the ratio

$$R^{\alpha}_{4+/2+} = \frac{E^{\alpha}_{4+} - E^{\alpha}_{0+}}{E^{\alpha}_{2+} - E^{\alpha}_{0+}} \tag{13.52}$$

may be deduced.

For ${}^{166}_{90}$Os a value of $R_{4+/2+} = 2.29$ and for ${}^{168}_{92}$Os a value of $R_{4+/2+} = 2.35$ results. These predictions are yet to be verified by experiment.

Furthermore, the sequence of osmium isotopes, presented in Fig. 13.8 indicates the position of the next magic neutron number $N = 126$. The nucleus ${}^{184}_{106}$Os with the maximum $\alpha$ value in the sequence of osmium isotopes, is positioned just between the two magic numbers $N = 82$ and $N = 126$ and may be used for an estimate for the next magic neutron number. Therefore, we propose an alternative approach to the search for super-heavy elements. Instead of a direct synthesis of super-heavy elements

[Oganessian(2004), Hofmann(2000), Oganessian(2010), Düllmann(2010)] a systematic survey of $\alpha$ values of the ground state band spectra of even-even nuclei with $N > 158$ and $Z > 98$ could reveal indirect information about the next expected magic numbers $Z = 114$ and $N = 184$. The currently available experimental data do not suffice to determine these magic numbers accurately.

From the results presented we draw the conclusion, that the full variety of low energy ground state spectra of even-even nuclei is described with a high grade of accuracy from a generalised point of view within the framework of the fractional symmetric rigid rotor.

The great advantage of the fractional symmetric rigid rotor model compared to the classical geometric models is, that nuclear ground state band spectra are described analytically with minimal effort according to equation (13.49) and with an excellent accuracy.

# Chapter 14

# q-deformed Lie Algebras and Fractional Calculus

*In the previous sections we have demonstrated, that the extension of the standard rotation group $SO(N)$, whose elements are the standard angular momentum operators to the fractional rotation group $SO^\alpha(N)$, whose elements are the fractional angular momentum operators, leads to extended symmetries. In this case the extension is from pure rotational symmetry to fractional rotation symmetry, which besides rotations also includes vibrational degrees of freedom.*

*The transition from a classical symmetry group to the analogue fractional symmetry group always implies a generalization of the classical symmetry and allows for a broader description of natural phenomena.*

*There is a different branch of physics which deals with exactly the same subject, too. This is the theory of q-deformed Lie algebras, which may be interpreted as extended versions of standard Lie groups.*

*The combination of concepts and methods developed in different branches of physics, has always led to new insights and improvements. The intention of this chapter is to show, that the concept of q-deformed Lie algebras and the methods developed in fractional calculus are strongly related (both are based on a generalized derivative definition) and may be combined leading to a new class of fractional deformed Lie algebras.*

*An interesting question of physical relevance is if deformed Lie algebras are not only suitable for describing small deviations from Lie symmetries, but in addition can bridge different Lie symmetries. We will demonstrate that fractional q-deformed Lie algebras show exactly this behaviour.*

## 14.1 q-deformed Lie algebras

q-deformed Lie algebras are extended versions of the usual Lie algebras [Bonatsos(1999)]. They provide us with an extended set of symmetries and therefore allow the description of physical phenomena, which are beyond the scope of usual Lie algebras. In order to describe a q-deformed Lie algebra, we introduce a parameter $q$ and define a mapping of ordinary numbers $x$ to q-numbers e.g. via:

$$[x]_q = \frac{q^x - q^{-x}}{q - q^{-1}} \tag{14.1}$$

which in the limit $q \to 1$ yields the ordinary numbers

$$\lim_{q \to 1} [x]_q = x \tag{14.2}$$

and furthermore we obtain as a special value

$$[0]_q = 0 \tag{14.3}$$

Based on q-numbers a q-derivative may be defined via:

$$D_x^q f(x) = \frac{f(qx) - f(q^{-1}x)}{(q - q^{-1})x} \tag{14.4}$$

With this definition for a function $f(x) = x^n$ we get

$$D_x^q \, x^n = [n]_q \, x^{n-1} \tag{14.5}$$

This is a first common aspect of q-deformation and fractional calculus: they both introduce a generalized derivative operator.

As an example for q-deformed Lie algebras we introduce the q-deformed harmonic oscillator. The creation and annihilation operators $a^\dagger$, $a$ and the number operator $N$ generate the algebra:

$$[N, a^\dagger] = a^\dagger \tag{14.6}$$

$$[N, a] = -a \tag{14.7}$$

$$aa^\dagger - q^{\pm 1} a^\dagger a = q^{\mp N} \tag{14.8}$$

with the definition (14.1) equation (14.8) may be rewritten as

$$a^\dagger a = [N]_q \tag{14.9}$$

$$aa^\dagger = [N+1]_q \tag{14.10}$$

Defining a vacuum state with $a|0\rangle = 0$, the action of the operators $\{a, a^\dagger, N\}$ on the basis $|n\rangle$ of a Fock space, defined by a repeated action of the creation operator on the vacuum state, is given by:

$$N|n\rangle = n|n\rangle \tag{14.11}$$

$$a^\dagger|n\rangle = \sqrt{[n+1]_q}|n+1\rangle \tag{14.12}$$

$$a|n\rangle = \sqrt{[n]_q}|n-1\rangle \tag{14.13}$$

The Hamiltonian of the q-deformed harmonic oscillator is defined as

$$H = \frac{\hbar\omega}{2}(aa^\dagger + a^\dagger a) \tag{14.14}$$

and the eigenvalues on the basis $|n\rangle$ result as

$$E^q(n) = \frac{\hbar\omega}{2}([n]_q + [n+1]_q) \tag{14.15}$$

There is a common aspect in the concepts of fractional calculus and q-deformed Lie algebras: they both extend the definition of the standard derivative operator.

Therefore we will try to establish a connection between the q-derivative (14.4) and the fractional derivative definition. To be conformal with the requirements (14.3) and (14.5), we will apply the Caputo derivative $D_x^\alpha$:

$$D_x^\alpha f(x) = \begin{cases} \frac{1}{\Gamma(1-\alpha)} \int_0^x d\xi\, (x-\xi)^{-\alpha} \frac{\partial}{\partial \xi} f(\xi) & 0 \le \alpha < 1 \\ \frac{1}{\Gamma(2-\alpha)} \int_0^x d\xi\, (x-\xi)^{1-\alpha} \frac{\partial^2}{\partial \xi^2} f(\xi) & 1 \le \alpha < 2 \end{cases} \tag{14.16}$$

For a function set $f(x) = x^{n\alpha}$ we obtain:

$$D_x^\alpha x^{n\alpha} = \begin{cases} \frac{\Gamma(1+n\alpha)}{\Gamma(1+(n-1)\alpha)} x^{(n-1)\alpha} & n > 0 \\ 0 & n = 0 \end{cases} \tag{14.17}$$

Let us now interpret the fractional derivative parameter $\alpha$ as a deformation parameter in the sense of q-deformed Lie algebras. Setting $|n\rangle = x^{n\alpha}$ we define:

$$[n]_\alpha |n\rangle = \begin{cases} \frac{\Gamma(1+n\alpha)}{\Gamma(1+(n-1)\alpha)} |n\rangle & n > 0 \\ 0 & n = 0 \end{cases} \tag{14.18}$$

Indeed it follows

$$\lim_{\alpha \to 1} [n]_\alpha = n \tag{14.19}$$

The more or less abstract q-number is now interpreted within the mathematical context of fractional calculus as the fractional derivative parameter $\alpha$ with a well-understood meaning.

The definition (14.18) looks just like one more definition for a q-deformation. But there is a significant difference which makes the definition based on fractional calculus unique.

Standard q-numbers are defined more or less heuristically. There exists no physical or mathematical framework, which determines their explicit structure. Consequently, many different definitions have been proposed in the literature (see e.g. [Bonatsos(1999)]).

In contrast to this diversity, the q-deformation based on the definition of the fractional derivative is uniquely determined, once a set of basis functions is given.

As an example we will derive the q-numbers for the fractional harmonic oscillator in the next section.

## 14.2 The fractional q-deformed harmonic oscillator

In the following we will present a derivation of a fractional q-number, which is independent of the specific choice of a fractional derivative definition.

The classical Hamilton function of the harmonic oscillator is given by

$$H_{\text{class}} = \frac{p^2}{2m} + \frac{1}{2} m\omega^2 x^2 \tag{14.20}$$

Following the canonical quantization procedure we replace the classical observables $\{x, p\}$ by the fractional derivative operators $\{\hat{X}, \hat{P}\}$ according to (9.10) and (9.11). The quantized Hamilton operator $H^\alpha$ results:

$$H^\alpha = \frac{\hat{P}^2}{2m} + \frac{1}{2} m\omega^2 \hat{X}^2 \tag{14.21}$$

The stationary Schrödinger equation is given by

$$H^\alpha \Psi = \left( -\frac{1}{2m} \left( \frac{\hbar}{mc} \right)^{2\alpha} m^2 c^2 \, D_x^\alpha D_x^\alpha + \frac{1}{2} m\omega^2 \left( \frac{\hbar}{mc} \right)^{2(1-\alpha)} x^{2\alpha} \right) \Psi = E\Psi \tag{14.22}$$

Introducing the variable $\xi$ and the scaled energy $E'$:

$$\xi^\alpha = \sqrt{\frac{m\omega}{\hbar}} \left( \frac{\hbar}{mc} \right)^{1-\alpha} x^\alpha \tag{14.23}$$

$$E = \hbar\omega E' \tag{14.24}$$

we obtain the stationary Schrödinger equation for the fractional harmonic oscillator in the canonical form

$$H^\alpha \Psi_n(\xi) = \frac{1}{2} (-D_\xi^{2\alpha} + |\xi|^{2\alpha}) \Psi_n(\xi) = E'(n, \alpha) \Psi_n(\xi) \tag{14.25}$$

In contrast to the classical harmonic oscillator this fractional Schrödinger equation has not been solved analytically until now. Since we are interested in an analytic expression for the energy levels, we will use the

Bohr–Sommerfeld quantization rule to obtain an analytic approximation [Laskin(2002)]:

$$2\pi\hbar\left(n+\frac{1}{2}\right)=\oint p d\xi \tag{14.26}$$

from (14.25) it follows for the momentum

$$|p|=\left(2E'-|\xi|^{2\alpha}\right)^{\frac{1}{2\alpha}} \tag{14.27}$$

For classical turning points the condition $p = 0$ holds. Motion in the classical sense therefore is allowed for $|\xi| \leq (2E')^{\frac{1}{2\alpha}}$. For (14.26) it follows explicitly:

$$2\pi\hbar\left(n+\frac{1}{2}\right)=4\int_0^{(2E')^{\frac{1}{2\alpha}}} p d\xi = 4\int_0^{(2E')^{\frac{1}{2\alpha}}}\left(2E'-|\xi|^{2\alpha}\right)^{\frac{1}{2\alpha}} d\xi \tag{14.28}$$

The integral may be solved:

$$E'(n,\alpha)=\left(\frac{1}{2}+n\right)^{\alpha}\pi^{\alpha/2}\left(\frac{\alpha\Gamma(\frac{1+\alpha}{2\alpha})}{\Gamma(\frac{1}{2\alpha})}\right)^{\alpha}, \quad n=0,1,2,... \tag{14.29}$$

This is the analytic approximation for the energy spectrum of the fractional harmonic oscillator. Furthermore this expression is independent of a specific choice of the fractional derivative definition.

In view of q-deformed Lie algebras, we can use this analytic result to derive the corresponding q-number. With (14.15) the q-number is determined by the recursion relation:

$$[n+1]_\alpha = 2E'(n,\alpha)-[n]_\alpha \tag{14.30}$$

In order to solve this equation, an appropriately chosen initial condition is necessary. The obvious choice $[0]_\alpha = 0$ for the initial condition leads to an oscillatory behaviour for $[n]_\alpha$ for $\alpha < 1$. If we require a monotonically increasing behaviour of $[n]_\alpha$ for increasing $n$, an adequate choice for the initial condition is

$$[0]_\alpha = 2^{1+\alpha}\pi^{\alpha/2}\left(\frac{\alpha\Gamma(\frac{1+\alpha}{2\alpha})}{\Gamma(\frac{1}{2\alpha})}\right)^{\alpha}\left(\zeta\left(-\alpha,\frac{1}{4}\right)-\zeta\left(-\alpha,\frac{3}{4}\right)\right) \tag{14.31}$$

The explicit solution is then given by:

$$[n]_\alpha = 2^{1+\alpha}\pi^{\alpha/2}\left(\frac{\alpha\Gamma(\frac{1+\alpha}{2\alpha})}{\Gamma(\frac{1}{2\alpha})}\right)^{\alpha}\left(\zeta\left(-\alpha,\frac{1}{4}+\frac{n}{2}\right)-\zeta\left(-\alpha,\frac{3}{4}+\frac{n}{2}\right)\right) \tag{14.32}$$

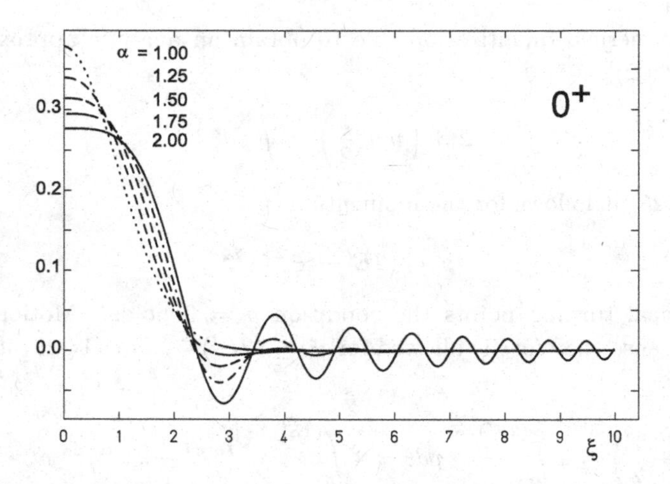


Fig. 14.1. Plot of the ground-state wave function $\Psi_{0^+}(\xi)$ of the fractional harmonic oscillator (14.25), solved numerically with the Caputo fractional derivative (14.16) for different $\alpha$. Numerical methods see e.g. [Diethelm(2005)].

where $\zeta(s,x)$ is the incomplete Riemann or Hurwitz zeta function, which is defined as:

$$\zeta(s,x) = \sum_{k=0}^{\infty}(k+x)^{-s} \tag{14.33}$$

and for $\alpha = m \in 0, 1, 2, ...$ it is related to the Bernoulli polynomials $B_m$ via:

$$\zeta(-m,x) = -\frac{1}{(m+1)}B_{m+1}(x) \tag{14.34}$$

Of course, the vacuum state $|0\rangle$ is no more characterized by a vanishing expectation value of the annihilation operator, but is defined by a zero expectation value of the number operator, which is the inverse function of the fractional q-number (14.32):

$$N|0\rangle = \left([n]_\alpha\right)^{-1}|0\rangle = n|0\rangle = 0 \tag{14.35}$$

Following an idea of Goldfain [Goldfain(2008a)] we may interpret the fractional derivative as a simultaneous description of a particle and a corresponding gauge field. Interpreting the vacuum state as a particles absent, but gauge field present state, the non-vanishing expectation value of the annihilation operator indicates the presence of the gauge field, while the number operator only counts for particles.

Setting $\alpha = 1$ leads to $[n]_{\alpha=1} = n$ and consequently, the standard harmonic oscillator energy spectrum $E'(n, \alpha = 1) = (1/2 + n)$ results. For $\alpha = 2$ it follows

$$E'(n, \alpha = 2) = 4\pi \left(\frac{\Gamma(\frac{3}{4})}{\Gamma(\frac{1}{4})}\right)^2 (\frac{1}{2} + n)^2 \tag{14.36}$$

$$= \frac{8\pi^3}{\Gamma(\frac{1}{4})^4} (\frac{1}{4} + n + n^2) \tag{14.37}$$

$$= \frac{2\pi^3}{\Gamma(\frac{1}{4})^4} + \frac{8\pi^3}{\Gamma(\frac{1}{4})^4} (n(n+1)) \tag{14.38}$$

which matches, besides a non vanishing zero point energy contribution, a spectrum of rotational type $E_{\text{rot}} \equiv l(l+1), l = 0, 1, 2, ...$ Unlike applications of ordinary q-numbers, this result is not restricted to a finite number $n$, but is valid for all $n$. This is a significant enhancement and it allows us to apply this model for high-energy excitations, too.

We obtained two idealized limits for the energy spectrum of the fractional harmonic oscillator. For $\alpha = 1$ a vibration type spectrum is generated, while for $\alpha = 2$ a rotational spectrum results. The fractional derivative coefficient $\alpha$ acts like an order parameter and allows for a smooth transition between these idealized limits.

Therefore the properties of the fractional harmonic oscillator seem to be well suited to describe e.g. the ground-state band spectra of even-even nuclei. We define

$$E_J^{\text{vib}}(\alpha, m_0^{\text{vib}}, a_0^{\text{vib}}) = m_0^{\text{vib}} + a_0^{\text{vib}}([J]_\alpha + [J+1]_\alpha)/2, \qquad J = 0, 2, 4, ... \tag{14.39}$$

where $m_0^{\text{vib}}$ mainly acts as a counter term for the zero-point energy and $a_0^{\text{vib}}$ is a measure for the level spacing.

Using (14.39) for a fit of the experimental ground-state band spectra of $^{156}$Gd, $^{196}$Pt, $^{110}$Cd and $^{218}$Po, which represent typical rotational-, $\gamma$-unstable-, vibrational and linear type spectra, in Table 14.1 the optimum parameter sets are listed. Below the onset of microscopic alignment effects all spectra are described with an accuracy better than 2% and are of similar accuracy like the results presented for the symmetric fractional rotor in the last chapter.

Therefore the fractional q-deformed harmonic oscillator indeed describes the full variety of ground-state bands of even-even nuclei with remarkable accuracy.

Table 14.1. Listed are the optimum parameter sets ($\alpha$, $a_0^{\text{vib}}$, $m_0^{\text{vib}}$ according to (14.39)) for the fractional harmonic oscillator for different nuclides. The maximum angular momentum $J_{\max}$ for a valid fit below the onset of alignment effects is given as well as the root mean square error $\Delta$E between experimental and fitted energies in %. The values may be compared with Table 13.1.

| nuclid | $\alpha$ | $a_0^{\text{vib}}[keV]$ | $m_0^{\text{vib}}[keV]$ | $J_{\max}$ | $\Delta$E[%] |
|---|---|---|---|---|---|
| ${}^{156}_{92}\text{Gd}_{64}$ | 1.795 | 15.736 | −14.136 | 14 | 1.48 |
| ${}^{196}_{118}\text{Pt}_{78}$ | 1.436 | 91.556 | −39.832 | 10 | 0.16 |
| ${}^{110}_{62}\text{Cd}_{48}$ | 1.331 | 197.119 | −87.416 | 6 | 0 |
| ${}^{218}_{134}\text{Po}_{84}$ | 0.801 | 357.493 | −193.868 | 8 | 0.06 |

The proposed fractional q-deformed harmonic oscillator is a fully analytic model, which may be applied with minimum effort not only to the limiting idealized cases of the pure rotor and pure vibrator respectively, but covers all intermediate configurations as well.

In the next section we will demonstrate the equivalence of the proposed fractional q-deformed harmonic oscillator and the corresponding fractional q-deformed symmetric rotor.

## 14.3 The fractional q-deformed symmetric rotor

In the previous section we have derived the q-number associated with the fractional harmonic oscillator. Interpreting equation (14.32) as a formal definition, the Casimir operator $C(\text{SU}_\alpha(2))$

$$C(\text{SU}_\alpha(2)) = [\text{J}]_\alpha [\text{J}+1]_\alpha\,, \quad \text{J} = 0,1,2,... \tag{14.40}$$

of the group $\text{SU}_\alpha(2)$ is determined. This group is generated by the operators $J_+$, $J_0$ and $J_-$, satisfying the commutation relations:

$$[J_0, J_\pm] = \pm J_\pm \tag{14.41}$$

$$[J_+, J_-] = [2J_0]_\alpha \tag{14.42}$$

Consequently we are able to define the fractional q-deformed symmetric rotor as

$$E_J^{\text{rot}}(\alpha, m_0^{\text{rot}}, a_0^{\text{rot}}) = m_0^{\text{rot}} + a_0^{\text{rot}}\,[J]_\alpha\,[J+1]_\alpha\,, \quad J = 0,2,4,... \tag{14.43}$$

where $m_0^{\text{rot}}$ mainly acts as a counter term for the zero-point energy and $a_0^{\text{rot}}$ is a measure for the level spacing.

For $\alpha = 1$, $E^{\text{rot}}$ reduces to $E^{\text{rot}} = m_0^{\text{rot}} + a_0^{\text{rot}} J(J+1)$, which is the spectrum of a symmetric rigid rotor. For $\alpha = 1/2$ we obtain:

$$\lim_{J\to\infty} (E_{J+2}^{\text{rot}} - E_J^{\text{rot}}) = a_0^{\text{rot}}\pi/2 = \text{const} \tag{14.44}$$

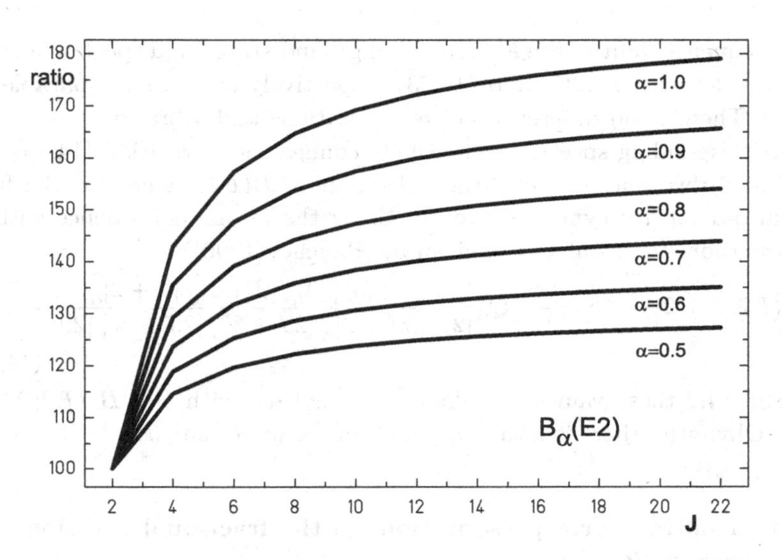


Fig. 14.2. B(E2)-values for the fractional q-deformed $SU_\alpha(2)$ symmetric rotor according to (14.50), normalized with $100/B_\alpha(E2; 2^+ \to 0^+)$.

which is the spectrum of a harmonic oscillator. We define the ratios

$$R^{\text{vib}}_{J,\alpha} = \frac{(E^{\text{vib}}_J - E^{\text{vib}}_0)}{(E^{\text{vib}}_2 - E^{\text{vib}}_0)} \tag{14.45}$$

$$R^{\text{rot}}_{J,\alpha} = \frac{(E^{\text{rot}}_J - E^{\text{rot}}_0)}{(E^{\text{rot}}_2 - E^{\text{rot}}_0)} \tag{14.46}$$

which only depend on $J$ and $\alpha$.

A Taylor series expansion at $J = 2$ and $\alpha = 1$ leads to:

$$\begin{aligned} R^{\text{vib}}_{J,\alpha} &= 1 + (J-2)(0.10 - 0.04(J-2)) \\ &\quad + (\alpha-1)(J-2)(0.101 - 0.020(J-2)) \end{aligned} \tag{14.47}$$

$$\begin{aligned} R^{\text{rot}}_{J,\alpha/2} &= 1 + (J-2)(0.11 - 0.04(J-2)) \\ &\quad + (\alpha-1)(J-2)(0.095 - 0.016(J-2)) \end{aligned} \tag{14.48}$$

A comparison of these series leads to the remarkable result

$$R^{\text{vib}}_{J,\alpha} \simeq R^{\text{rot}}_{J,\alpha/2} + o(J^3, \alpha^2) \tag{14.49}$$

Therefore the fractional q-deformed harmonic oscillator (14.39) and the fractional q-deformed symmetric rotor (14.43) generate similar spectra. As a consequence, a fit of the experimental ground-state band spectra of even-even nuclei with (14.39) and (14.43) respectively leads to comparable results. There is no difference between rotations and vibrations any more, the corresponding spectra are mutually connected via relation (14.49).

Finally we may consider the behaviour of $B(E2)$-values for the fractional q-deformed symmetric rotor. Using the formal equivalence with q-deformation, these values are given by [Raychev(1990)]:

$$B_\alpha(E2; J+2 \to J) = \frac{5}{16\pi} Q_0^2 \frac{[3]_\alpha [4]_\alpha [J+1]_\alpha^2 [J+2]_\alpha^2}{[2]_\alpha [2J+2]_\alpha [2J+3]_\alpha [2J+4]_\alpha [2J+5]_\alpha} \tag{14.50}$$

In Fig. 14.2 these values are plotted, normalized with $100/B_\alpha(E2; 2^+ \to 0^+)$. Obviously there is a saturation effect for increasing $J$.

## 14.4 Half-integer representations of the fractional rotation group $SO^\alpha(3)$

Up to now we have investigated the ground-state excitation spectra of even-even nuclei. The next step of complication arises if the proton or neutron number is odd. In classical models it is expected, that excitations of both collective and single-particle type will be possible and in general will be coupled. Approximately, the even-even nucleus is treated as a collective core, whose internal structure is not affected by one more particle moving on its surface. In the strong-coupling model, the corresponding strong coupling Hamiltonian $H_{sc}$ is decomposed into a collective, single-particle and interaction term [Eisenberg(1987)]:

$$H_{\text{sc}} = H^0_{\text{coll}} + H^0_{\text{sp}} + H_{\text{ii}} \tag{14.51}$$

For $K = 1/2$ ground-state bands the energy level spectrum is known analytically [Eisenberg(1987)]:

$$E_{K=1/2}(I) = m_0 + c_0 I(I+1) + a_0(-1)^{I+1/2}(I+1/2), \quad I = 1/2, 3/2, 5/2, ... \tag{14.52}$$

where $a_0$ is called the decoupling parameter and $m_0$ and $c_0$ in units (keV) are parameters to be fitted with experimental data.

In view of a q-deformed Lie algebra of the standard rotation group $SO(3)$ this result may be interpreted as an expansion in terms of of Casimir operators:

$$E_{K=1/2}(I) = m_0 + c_0 C(SO(3)_q) + a_0(-1)^{I+1/2} C(SO(2)_q) \tag{14.53}$$
$$= m_0 + c_0 [I]_q [I+1]_q + a_0(-1)^{I+1/2} [I+1/2]_q \tag{14.54}$$

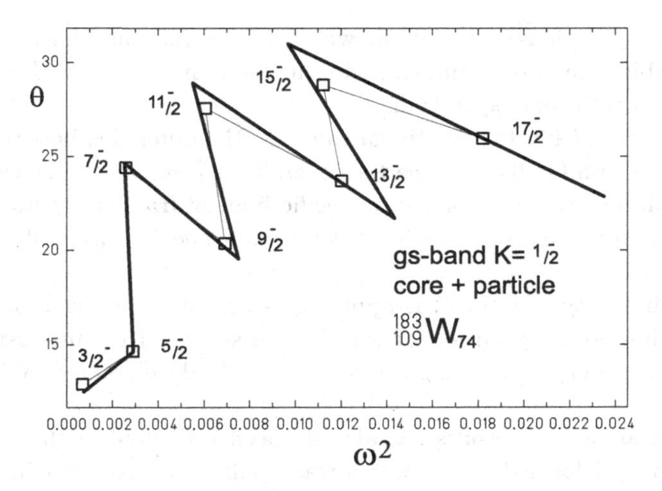


Fig. 14.3. Back-bending plot for $^{183}W_{74}$ for the $K = 1/2^-$ ground-state band. Thick line is the fit result according to (14.56). Squares indicate the experimental values.

Consequently the investigation of the ground-state band spectra of even-odd and odd-even nuclei respectively is a test of whether or not half-integer representations of the fractional q-deformed rotation group are realized in nature.

Furthermore, in view of fractional calculus, we may use the explicit form (14.32) of the q-number associated with the fractional harmonic oscillator:

$$E_{K=1/2}(I) = m_0 + c_0 C(SO(3)_\alpha) + a_0(-1)^{I+1/2} C(SO(2)_\alpha) \quad (14.55)$$
$$= m_0 + c_0[I]_\alpha[I+1]_\alpha + a_0(-1)^{I+1/2}[I+1/2]_\alpha \quad (14.56)$$

In Fig. 14.3 we show a fit of the experimental values for the $K = 1/2^-$ band of the nucleus $^{183}W_{74}$ with (14.56). With $\alpha = 1.02$, $m_0 = -6.077$, $c_0 = 11.76$ and $a_0 = 2.09$ the rms-error is less than 1%. This is a first indication, that half-integer representations of the fractional rotation group may be indeed realized in nature.

In this chapter we have demonstrated, that we are able to describe the strong relation between rotations and vibrations on the basis of two different approaches.

On one hand we presented the fractional symmetric rotor which contains as the classical limit $\alpha = 1$ the standard rotation spectrum and generates a vibrational spectrum for the semiderivative $\alpha = \frac{1}{2}$.

On the other hand we have shown, that the fractional harmonic oscillator exhibits the vibrational spectrum in the limit $\alpha = 1$, while for $\alpha = 2$ we obtain a rotational spectrum.

Therefore, within the fractional calculus, the difference between $SO(3)$ and $U(1)$ is annihilated. Instead a generalized view is established, which allows to interpret vibrations as a specific limit of fractional rotations and vice versa, rotations may be interpreted as a specific limit of fractional vibrations.

The similarity of intentions applying q-deformed Lie algebras and the use of a fractional extension of well known standard groups respectively may also be interpreted as a guide for the future development of fractional calculus.

A vast amount of results has already been presented in the past. The concept of q-deformed Lie algebras was applied to problems in different branches of physics.

A possible extension of these research areas within the framework of a fractional group theory could lead to a broader understanding of both concepts and could provide new insights into the structure and variety of extended symmetries in nature.

# Chapter 15

# Fractional Spectroscopy of Hadrons

*The calculation of the hadron spectrum is a task of central interest in quantum chromodynamics (QCD), which describes the dynamics of particles which are subject to strong interaction forces. This requires enormous efforts and computer power [Dürr(2008)], that is one reason, why different approximate strategies have been established, which reduce the complexity of the problem.*

*Most prominent representatives are non-relativistic potential models, which are very easy to handle. A standard Schrödinger equation is solved with an appropriately chosen model potential. Minimum requirements for such potentials are, that for small distances the one-gluon exchange potential should be simulated while for large distances the quark confinement should be modelled.*

*A simple ansatz is given with a rotational symmetric potential [Eichten(1975)]*

$$V(r) = -\frac{\bar{\alpha}_s}{r} + \kappa r \tag{15.1}$$

*where a running coupling constant* $\bar{\alpha}_s$ *and a force constant* $\kappa$ *are parameters of this model. For a realistic treatment of fine- and hyperfine-structure of spectra it is necessary, to introduce additional spin-spin, spin-orbit and tensor interaction terms [Godfrey(1985), Barnes(1995), Li(2009)].*

*Such potential models describe the low energy excitation spectrum of meson and baryon spectra with reasonable accuracy. But in the meantime more and more recently discovered particles do not fit very well into the calculated term schemes [Zhu(2005), Stancu(2010), Pak(2010)].*

*In this section we want to demonstrate that fractional calculus provides excellent contributions for a successful description and understanding of hadron spectra.*

*We want to remind, that the solutions of a free fractional Schrödinger equation, presented in Chapter 9 are localized in a very small region of space. A direct consequence of this behaviour may be quark confinement. Furthermore we have demonstrated, that this behaviour of free, fractional and localized solutions of a fractional Schrödinger equation may be simulated using a standard Schrödinger equation with a linear potential.*

*Therefore we assume, that the experimental hadron spectra may be successfully be described with a free fractional Schrödinger equation. A specific additional potential term is not necessary.*

*That this hope is justified will be shown in the next sections.*

## 15.1 Phenomenology of the baryon spectrum

In order to classify the multiplets of the fractional rotation group $SO^{\alpha}(3)$ we derived the group chain

$$SO^{\alpha}(3) \supset SO^{\alpha}(2) \tag{15.2}$$

In the last section we presented a successful interpretation of low energy excitation spectra for even-even nuclei based of the fractional symmetric rotor model, which is based on the Casimir operator $L^{\alpha\,2}$ of $SO^{\alpha}(3)$. Until now we have ignored a possible influence of the Casimir operator $L_z^{\alpha}$ of $SO^{\alpha}(2)$. We start with a model Hamilton operator $H^{\alpha}$

$$H^{\alpha} = m_0 + a_0 L^{\alpha\,2} + b_0 L_z^{\alpha} \tag{15.3}$$

where coefficients $m_0$, $a_0$ and $b_0$ are free parameters. A complete classification of multiplets of this operator may be realized, introducing the Casimir operators (12.45) and (12.46). The levels are uniquely determined for a given set of $L$ and $M$:

$$E_R^{\alpha} = m_0 + a_0 \frac{\Gamma(1+(L+1)\alpha)}{\Gamma(1+(L-1)\alpha)} \pm b_0 \frac{\Gamma(1+|M|\alpha)}{\Gamma(1+(|M|-1)\alpha)} \tag{15.4}$$

Note that for $\alpha = 1$ the Casimir operators and the corresponding eigenvalues reduce to the well known results of standard quantum mechanical angular momentum algebra [Edmonds(1957)].

$$E_R^{\alpha=1} = m_0 + a_0\, L(L+1) + b_0\, M \tag{15.5}$$

Let us recall, that the motion of an electron around the atomic nucleus under the influence of a constant magnetic field $B_z$, which is known as normal Zeeman effect [Zeeman(1897)] ignoring a spin dependency is described in

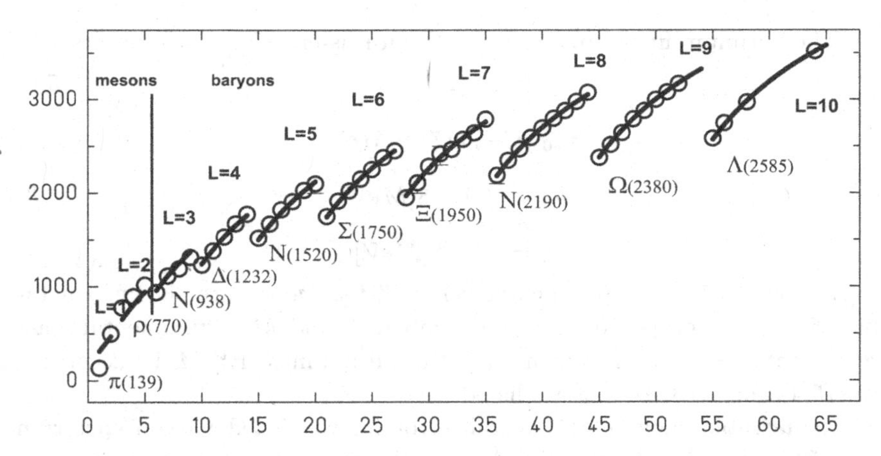


Fig. 15.1. A comparison of the experimental baryon spectrum [Nakamura(2010)] with a fit with $E_R^\alpha$ from (15.4), which generates the full spectrum of the fractional rotation group $SO^\alpha(3)$. Circles denote the experimental mass, theoretical values are drawn as lines for a given $L$. Band heads are labelled with the corresponding particle name and mass. For $L = 1, 2$ the calculated predictions for baryon masses are compared with the masses of mesons.

this way as a dependence on the angular momentum $L$ and its projection onto the z-axis $M$. We are led to the conclusion, that parameter $b_0$ may be interpreted as the fractional analogue to a magnetic field.

Hence we interpret (15.4) as a description of a non-relativistic spin-0 particle in constant fractional magnetic field with strength $b_0$.

If we associate this particle with a quark and the fractional magnetic field with a colour magnetic field, we obtain a simple analytic fractional model, which should allow a description of the hadron spectrum.

This assumption is motivated by results obtained by ab initio calculations in lattice QCD [Wilson(1974)], where a colour flux tube with almost constant colour magnetic field between quarks is observed [Casher(1979)].

Since we expect a similar result for self consistent solutions of the non-linear QCD field equations, the fractional Zeeman-effect serves as a simple, idealized model for a first test, whether the hadron spectrum may be reproduced at all within the framework of a fractional Abelian field theory.

Therefore we will use formula (15.4) for a fit of the experimental baryon spectrum [Nakamura(2010)].

The optimum fit parameter set with an rms-error of 0.84% results as:

$$\alpha = 0.112 \tag{15.6}$$

$$m_0 = -17171.6[MeV] \tag{15.7}$$

$$a_0 = 10971.8[MeV] \tag{15.8}$$

$$b_0 = 8064.6[MeV] \tag{15.9}$$

In Fig. 15.1 the experimental versus fitted values are shown. In table 15.1 name, proposed quantum numbers $L$ and $M$, experimental mass $E^{\text{exp}}[MeV]$ taken from [Nakamura(2010)], fitted mass $E^{\alpha}_{R}[MeV]$ according to (15.4) and error $\Delta\text{E}[\%]$ are listed.

Remarkable enough, the overall error for the fitted baryon spectrum $L = 3, \ldots, 9$ is less than 1 %. A general trend is an increasing error for small masses which possibly indicates the limitations of a non-relativistic model.

For the fractional symmetric rigid rotor we have demonstrated in Sec. 13.5 that the limit $\alpha = 0.11$ corresponds in a geometric picture to the case of a linear potential. Therefore the optimum fit value for $\alpha = 0.112$ represents the correct linear type for the long range potential used in phenomenological quark models.

In addition, the $\alpha = 0.112$ value corresponds in the sense of the relativistic theory (see Chapter 11) to an inherent $SU(18)$ symmetry, which optimistically may be interpreted as the direct product:

$$SU(18) = SU(N = 6)^{\text{flavour}} \otimes SU(3)^{\text{colour}} \tag{15.10}$$

and is therefore an indirect indication for the maximum number of different quark flavours to be found in nature.

In the lower part of Table 15.1 the proposed $(L, M)$ values for $\Lambda_c(2880)$ up to $\Lambda_b$ give a rough estimate for the number of particles, which are still missing in the experimental spectrum.

For $L = 1, 2$ we can compare the theoretical mass predictions for baryons with the experimental meson spectrum, see upper part of Table 15.1 and left part of Fig. 15.1. The results are surprisingly close to the experimental masses, despite the fact that mesons and baryons are different constructs. This means, that the relative strength of the fractional magnetic field $b_0 = B_q/m_q$, $\{q \in u, d, s, c\}$ is of similar magnitude in mesonic $(q\bar{q})$ and baryonic $qqq$ systems.

Summarizing the results of this section we conclude, that the fractional Zeeman effect, which describes the motion of a fractional charged particle

Table 15.1. A comparison of the experimental baryon spectrum with the mass formula (15.4). Listed are name, proposed quantum numbers $L$ and $M$, experimental mass $E^{\exp}[MeV]$, fitted mass $E_R^{\alpha}[MeV]$ and error $\Delta$E[%]. For $L = 1$ and $L = 2$ the predicted baryon masses are compared with experimental meson masses. * indicates the status of a particle.

| name | L | M | $E^{\exp}$[MeV] | $E_R^{\alpha}$[MeV] | $\Delta$E[%] |
|---|---|---|---|---|---|
| $\pi^0$ | 1 | 0 | 135 | 313.90 | 132.57 |
| $K_s^0$ | 1 | 1 | 498 | 470.45 | −5.46 |
| $\rho(770)$ | 2 | 0 | 776 | 652.41 | −15.87 |
| $K^*(892)^0$ | 2 | 1 | 896 | 808.96 | −9.71 |
| $\phi(1020)$ | 2 | 2 | 1019 | 945.76 | −7.19 |
| $N$ | 3 | 0 | 938 | 959.39 | 2.28 |
| $\Lambda$ | 3 | 1 | 1116 | 1115.94 | 0.02 |
| $\Sigma^0$ | 3 | 2 | 1193 | 1252.74 | 5.04 |
| $\Xi^0$ | 3 | 3 | 1315 | 1374.16 | 4.51 |
| $\Delta(1232)$ | 4 | 0 | 1232 | 1240.53 | 0.69 |
| $\Sigma^0(1385)$ | 4 | 1 | 1384 | 1397.08 | 0.97 |
| $\Xi(1530)$ | 4 | 2 | 1532 | 1533.87 | 0.14 |
| $\Omega^-$ | 4 | 3 | 1672 | 1655.29 | −1.00 |
| $\Lambda(1800)$ | 4 | 4 | 1775 | 1764.44 | −0.60 |
| $\Lambda(1520)$ | 5 | 0 | 1520 | 1500.10 | −1.28 |
| $\Lambda(1670)$ | 5 | 1 | 1670 | 1656.65 | −0.80 |
| $\Xi(1820)$ | 5 | 2 | 1823 | 1793.45 | −1.62 |
| $\Delta(1910)$ | 5 | 3 | 1910 | 1914.87 | 0.25 |
| $\Sigma(2030)$ | 5 | 4 | 2030 | 2024.01 | −0.29 |
| $\Lambda(2100)$ | 5 | 5 | 2100 | 2123.14 | 1.10 |
| $\Sigma(1750)$ | 6 | 0 | 1750 | 1741.42 | −0.49 |
| $\Sigma(1915)$ | 6 | 1 | 1915 | 1897.97 | −0.89 |
| $\Xi(2030)$ | 6 | 2 | 2025 | 2034.76 | 0.48 |
| $\Delta(2150)(*)$ | 6 | 3 | 2150 | 2156.18 | 0.29 |
| $\Omega(2250)$ | 6 | 4 | 2252 | 2265.33 | 0.59 |
| $\Omega(2380)(**)$ | 6 | 5 | 2380 | 2364.46 | −0.65 |
| $\Sigma_c(2452)$ | 6 | 6 | 2452 | 2455.27 | 0.13 |
| $\Xi(1950)$ | 7 | 0 | 1950 | 1967.06 | 0.88 |
| $\Lambda(2110)$ | 7 | 1 | 2110 | 2123.61 | 0.65 |
| $\Lambda_c$ | 7 | 2 | 2286 | 2260.41 | −1.14 |
| $\Delta(2420)$ | 7 | 3 | 2420 | 2381.83 | −1.58 |
| $\Xi_c^+(2467)$ | 7 | 4 | 2467 | 2490.97 | 0.97 |
| $\Xi_c^{'+}(2575)$ | 7 | 5 | 2576 | 2590.10 | 0.56 |
| $\Xi_c^0(2645)$ | 7 | 6 | 2645 | 2680.92 | 1.35 |
| $\Xi_c^0(2790)$ | 7 | 7 | 2791 | 2764.73 | −0.94 |

Table 15.1. (*Continued*)

| name | L | M | $E^{\exp}$[MeV] | $E_R^{\alpha}$[MeV] | ΔE[%] |
|---|---|---|---|---|---|
| $N(2190)$ | 8 | 0 | 2190 | 2179.12 | −0.50 |
| $\Lambda(2350)$ | 8 | 1 | 2350 | 2335.67 | −0.61 |
| $\Xi_c^0(2471)$ | 8 | 2 | 2471 | 2472.47 | 0.06 |
| $\Lambda_c^+(2593)$ | 8 | 3 | 2595 | 2593.89 | −0.06 |
| $\Omega_c^0$ | 8 | 4 | 2698 | 2703.03 | 0.21 |
| $\Sigma_c^0(2800)$ | 8 | 5 | 2800 | 2802.16 | 0.08 |
| $\Lambda_c^+(2880)$(***) | 8 | 6 | 2882 | 2892.98 | 0.38 |
| $\Xi(2980)$(***) | 8 | 7 | 2978 | 2976.79 | −0.06 |
| $\Xi_c(3080)$(***) | 8 | 8 | 3076 | 3054.61 | −0.70 |
| $\Omega^-(2380)$(**) | 9 | 0 | 2380 | 2379.27 | −0.03 |
| $\Sigma_c(2520)$(***) | 9 | 1 | 2518 | 2535.82 | 0.69 |
| $\Sigma_c(2645)$(***) | 9 | 2 | 2646 | 2672.62 | 1.00 |
| $\Xi_c^0$(***) | 9 | 3 | 2792 | 2794.04 | 0.20 |
| $\Lambda_c(2880)$(***) | 9 | 4 | 2882 | 2903.18 | 0.74 |
| $\Sigma(3000)$(*) | 9 | 5 | 3000 | 3002.31 | 0.10 |
| $\Xi_c^0$ (***) | 9 | 6 | 3080 | 3093.13 | 0.43 |
| $\Sigma(3170)$(*) | 9 | 7 | 3170 | 3176.94 | 0.22 |
| | 9 | 8 | | 3254 | |
| | 9 | 9 | | 3327 | |
| $\Lambda(2585)$(**) | 10 | 0 | 2585 | 2568.9 | 0.66 |
| $\Delta(2750)$(**) | 10 | 1 | 2750 | 2725 | 0.92 |
| | 10 | 2 | | 2862 | |
| $\Xi_c^0$ (***) | 10 | 3 | 2971 | 2983 | 0.40 |
| | 10 | 4 | | 3092 | |
| | 10 | 5 | | 3191 | |
| | 10 | 6 | | 3282 | |
| | 10 | 7 | | 3366 | |
| | 10 | 8 | | 3444 | |
| $\Xi_{cc}^+$(*) | 10 | 9 | 3519 | 3517.06 | −0.05 |
| $\Omega_{cc}$ | 11 | 8 | 3637 | 3627.63 | |
| $\Omega_{ccc}$ | 15 | 15 | 4681 | 4702.20 | |
| $\Lambda_b^0$(***) | 20 | 20 | 5620 | 5612.75 | −0.13 |

in a constant fractional magnetic field serves as a reasonable non-relativistic model for an understanding of the full baryon spectrum.

This result has far reaching consequences: For the first time the terms "particle described by a fractional wave equation" and "fractional magnetic field" have successfully been associated with the entities quark, which follows a $SU(n)^{\text{flavour}}$ symmetry and colour, which follows a $SU(3)^{\text{colour}}$ symmetry.

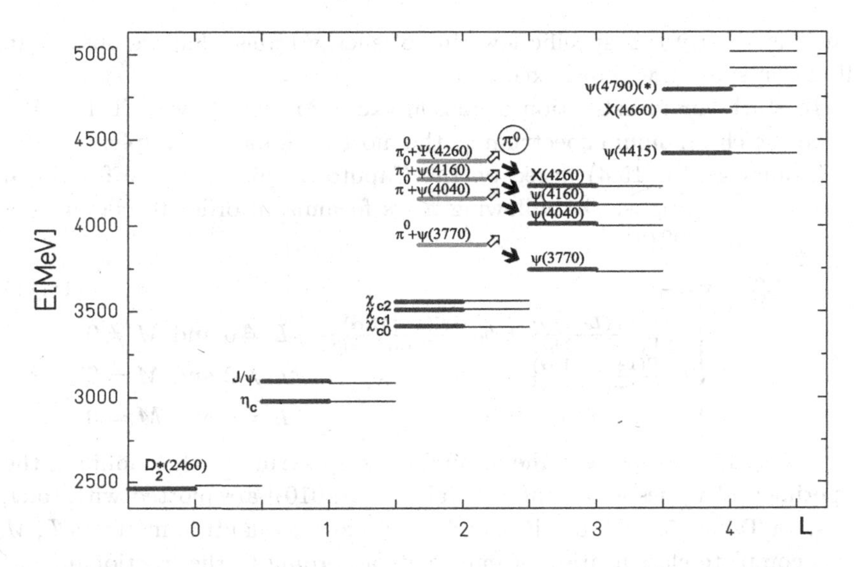


Fig. 15.2. Ground state bands of the charmonium spectrum. The experimental values are given as thick lines and theoretical values as small lines. Particles are labelled according to the quantum numbers of the fractional rotation group $SO^{\alpha}(3)$. While the overall agreement is very good, only for the multiplet $L = 3$ the theoretical predictions deviate by 133 [MeV]. Therefore we propose the following production process: First an excited short lived hybrid of type $u\bar{u}c\bar{c}^*$ and $d\bar{d}c\bar{c}^*$ respectively is produced, which decays into a pion and a long lived experimentally observed final state. This is indicated by gray lines for the theoretical mass of the predicted short lived hybrid state, arrows sketch the decay process.

## 15.2 Charmonium

So far, the presented results lead to the conclusion, that a fractional rotation group $SO^{\alpha}(3)$ based on the Riemann fractional derivative definition determines an inherent symmetry of the baryon excitation spectrum.

From this encouraging result immediately follows the question, whether the same symmetry is realized for mesonic excitation spectra too. In this section, we will present explicit evidence for that suggestion. But there is one fundamental difference: meson excitation spectra will be reproduced with an excellent degree of precision, if we apply the Caputo definition of the fractional derivative.

As a consequence, for $M = 0$ and $L = 0$ there will be a reduction of the corresponding energy levels in comparison with the levels obtained with the Riemann definition, while all other energy levels there will be no difference.

But it is exactly this specific lowering of energy levels, that we observe in all meson spectra without exception.

To start our investigation of meson excitation levels, we will first describe the charmonium spectrum as the most prominent example:

In analogy to (15.4) based on the Caputo definition of the fractional derivative we propose the following mass formula in order to classify the multiplets of $SO^\alpha(3)$:

$$E^\alpha_C = m_0 + \begin{cases} a_0 \frac{\Gamma(1+(L+1)\alpha)}{\Gamma(1+(L-1)\alpha)} \pm b_0 \frac{\Gamma(1+|M|\alpha)}{\Gamma(1+(|M|-1)\alpha)} & L \neq 0 \text{ and } M \neq 0 \\ a_0 \frac{\Gamma(1+(L+1)\alpha)}{\Gamma(1+(L-1)\alpha)} & L \neq 0 \text{ and } M = 0 \\ 0 & L = 0 \text{ and } M = 0 \end{cases} \tag{15.11}$$

In Fig. 15.2 we present the ground state spectrum of charmonium, the experimental values according to [Nakamura(2010)] are plotted with thick lines. In Table 15.2 these values and the proposed quantum numbers $L, M$ for a complete classification of multiplets according to the fractional rotation group are given.

We want to emphasize, that the presentation for reasons of simplicity is restricted to the ground state excitation spectrum of charmonium, higher order excitations like $\eta_c'$ will be associated with higher radial excitations and therefore will be ignored in a first attempt.

For a given $L = \text{const}$ multiplet the mass formula reduces to:

$$E^\alpha_C(\tilde{m}_0, b_0, \alpha) = \tilde{m}_0 \pm \begin{cases} b_0 \frac{\Gamma(1+|M|\alpha)}{\Gamma(1+(|M|-1)\alpha)} & M \neq 0 \\ 0 & M = 0 \end{cases} \tag{15.12}$$

with three parameters $\tilde{m}_0$, $b_0$ and $\alpha$. Let us apply (15.12) for $L = 2$ to the $\chi_c$ triplet, we have three determining equations for the parameters. Therefore we obtain:

$$\tilde{m}_0 = \chi_{c0} = 3414.75[\text{MeV}] \tag{15.13}$$

$$b_0 = 106.03[\text{MeV}] \tag{15.14}$$

$$\alpha = 0.677 \tag{15.15}$$

Hence we are confronted with a first encouraging result: The fractional derivative parameter $\alpha$ obtained from experimental data is very close to $\alpha = \frac{2}{3}$. From our derivation of a fractional Dirac equation we already found, that $\alpha = \frac{2}{n}$, which results from a n-fold factorization of the Klein–Gordon equation implies the existence of an inherent $SU(n)$ symmetry of the corresponding fractional wave equation.

Table 15.2. Comparison of experimental and theoretical masses according to (15.11) for the ground state band of charmonium. The values above $|41\rangle$ are predictions of the theoretical model.

| $\|LM\rangle$ | Name | $m_{\text{exp}}$ | $m_{\text{th}}$ |
|---|---|---|---|
| $\|00\rangle$ | $D_2^*(2460)$ | $2461.10 \pm 1.6$ | 2461.63 |
| $\|10\rangle$ | $\eta_c$ | $2980.5 \pm 1.2$ | 2989.58 |
| $\|11\rangle$ | $J/\psi$ | $3096.916 \pm 0.011$ | 3086.63 |
| $\|20\rangle$ | $\chi_{c0}$ | $3414.75 \pm 0.31$ | 3419.65 |
| $\|21\rangle$ | $\chi_{c1}$ | $3510.66 \pm 0.07$ | 3516.70 |
| $\|22\rangle$ | $\chi_{c2}$ | $3556.20 \pm 0.09$ | 3559.16 |
| $\|30\rangle$ | $\psi(3770)$ | $3772.92 \pm 0.35$ | 3763.12 |
| $\|31\rangle$ | $\psi(4040)$ | $4039 \pm 1$ | 4034.12 |
| $\|32\rangle$ | $\psi(4160)$ | $4153 \pm 3$ | 4152.68 |
| $\|33\rangle$ | $X(4260)$ | $4263 \pm 9$ | 4254.87 |
| $\|40\rangle$ | $\psi(4415)$ | $4421 \pm 4$ | 4413.10 |
| $\|41\rangle$ | $X(4660)$ | $4664 \pm 11 \pm 5$ | 4684.09 |
| $\|42\rangle$ | $\psi(4790)(*)$ | 4790 | 4802.66 |
| $\|43\rangle$ | $X$ | - | 4904.85 |
| $\|44\rangle$ | $X$ | - | 4996.81 |
| $\|50\rangle$ | $X$ | - | 4961.84 |
| $\|51\rangle$ | $X$ | - | 5232.84 |
| $\|52\rangle$ | $X$ | - | 5351.40 |
| $\|53\rangle$ | $X$ | - | 5453.59 |
| $\|54\rangle$ | $X$ | - | 5545.56 |
| $\|55\rangle$ | $X$ | - | 5630.34 |
| $\|60\rangle$ | $X$ | - | 5539.31 |
| $\|61\rangle$ | $X$ | - | 5810.30 |
| $\|62\rangle$ | $X$ | - | 5928.86 |
| $\|63\rangle$ | $X$ | - | 6031.05 |
| $\|64\rangle$ | $X$ | - | 6123.02 |
| $\|65\rangle$ | $X$ | - | 6207.80 |
| $\|66\rangle$ | $X$ | - | 6287.16 |

Furthermore for $L = 3$ we interpret the $\psi$ triplet as state $|30\rangle$, $|31\rangle$ and $|32\rangle$ and obtain another set of parameters:

$$\tilde{m}_0 = \psi(3770)|30\rangle = 3775.2[\text{MeV}] \tag{15.16}$$

$$b_0 = 293.8[\text{MeV}] \tag{15.17}$$

$$\alpha = 0.644 \tag{15.18}$$

This triplet is located above the $D - D^*$ meson threshold.

Within experimental errors, both $\alpha$ values are identical. This observation supports the assumption, that the spectrum may be interpreted using one unique $\alpha$. The difference between both triplets is mainly absorbed by the magnetic field strength parameter $b_0$, which suddenly increases by a factor 2.8. This may be directly related to the observation of a similar change of magnitude in the ratio R of cross-sections

$$R = \frac{\sigma(e^+e^- \to \text{Hadrons})}{\sigma(e^+e^-\mu^+\mu^-)} \tag{15.19}$$

which increases from $R = 2$ to $R = 5$ in this energy region.

The complete charmonium spectrum is therefore described by two distinct parameter sets below and above the threshold respectively:

$$E_C^\alpha = \begin{cases} E_C^\alpha(m_0, b_0^1, \alpha) & E < \text{threshold} \\ E_C^\alpha(m_0, b_0^2, \alpha) & E > \text{threshold} \end{cases} \tag{15.20}$$

Therefore we will apply $b_0^1$ and $b_0^2$ as two distinct fractional magnetic field strength parameters below and above the threshold respectively.

In addition, if our assumption of an inherent fractional rotation symmetry of the charmonium spectrum is valid, the experimentally observed triplet of $\psi$ particles is not complete, because the state $|33\rangle$ is missing.

Using the parameter set (15.16) we are able to predict the mass of the $|33\rangle$ state, which turns out to be in excellent agreement with the recently observed $X(4260)$ (formerly known as $Y(4260)$ [Aubert(2005)]) particle:

$$X(4260)_{\text{th}} = \psi|33\rangle = 4252.04[\text{MeV}] \tag{15.21}$$

This result also indicates, that the fractional magnetic field for a given $L =$ const multiplet is indeed a constant. Furthermore the proposed fractional model based on the fractional symmetry group $SO^\alpha(3)$ is the first one to give a reasonable classification for $X(4260)$ as a member of the $L = 3$ quartet.

Until now all standard models making use of phenomenological model potentials failed to explain the existence of $X(4260)$. This failure also inspired several investigations to interpret $X(4260)$ not as a simple charmonium excitation, but e.g. as a hybrid state [Zhu(2005)].

A complete fit of experimental data for the charmonium spectrum using (15.20) leads to a surprising result: The full spectrum is reproduced with excellent accuracy, but there is one exception: The theoretical masses for members of the multiplet $L = 3$ deviate from the experimental values by an amount of 133[MeV]. This is very close to the rest mass of the pion $\pi^0$.

We therefore suppose, that all members of the particle family $\psi(3770)$, $\psi(4040)$, $\psi(4160)$ and $X(4260)$ are generated via the same mechanism:

In a first step, a short lived hybrid state is formed e.g. $(u\bar{u}c\bar{c}^*)$ and $(d\bar{d}c\bar{c}^*)$ respectively: The mass of this hybrid state is exactly predicted by our model (15.11). Within a very short period of time this excited state decays into a pion and a long lived excited final state, which is observed experimentally:

$$X(3905) \rightarrow \pi^0 + \psi(3770) \tag{15.22}$$

$$X(4173) \rightarrow \pi^0 + \psi(4040) \tag{15.23}$$

$$X(4287) \rightarrow \pi^0 + \psi(4160) \tag{15.24}$$

$$X(4393) \rightarrow \pi^0 + X(4260) \tag{15.25}$$

Using this hypothesis, which yet has to be verified experimentally with the parameters

$$m_0 = 2461.63[\text{MeV}] \tag{15.26}$$

$$a_0 = 453.60[\text{MeV}] \tag{15.27}$$

$$b_0^1 = 107.85[\text{MeV}] \tag{15.28}$$

$$b_0^2 = 301.17[\text{MeV}] \tag{15.29}$$

$$\alpha = 0.648 \tag{15.30}$$

we calculate the theoretical masses, which are listed in Table 15.2 and agree with the experiment to a very high degree. Especially interesting is the state $|00\rangle$, which is not included within the framework of conventional potential models. We predict a particle with mass $m_0 = 2461.76[\text{MeV}]$. Indeed there exists a possible candidate in this energy region, which is $D_2^*(2460)$ a charmed meson.

Hence the symmetry of the fractional rotation group covers not only pure $c\bar{c}$ candidates but includes other charmed mesons as well.

In September 2007 the Belle collaboration, an international team of researchers at the High Energy Accelerator Research organization (KEK) in Tsukuba, Japan, found a new resonance at $X(4660)$, which may be directly identified as the $|41\rangle$-state of a fractional multiplet. This assignment also supports the assumption, that states above the meson-threshold may be characterized using a single unique magnetic field parameter $b_0^2$.

Figure 15.3 shows the results of the experiments performed by the Belle collaboration. The $X(4660)$ peak may be easily recognized. With arrows we marked additional positions of theoretically predicted peaks, which are also listed in the lower part of Table 15.2.

There is increasing evidence for a resonance $\psi(5S)$ at 4790[MeV] [Beveren(2009), Segovia(2010)], which could be interpreted as a realization of the $|42\rangle$-state with a predicted mass of 4802.66[MeV].

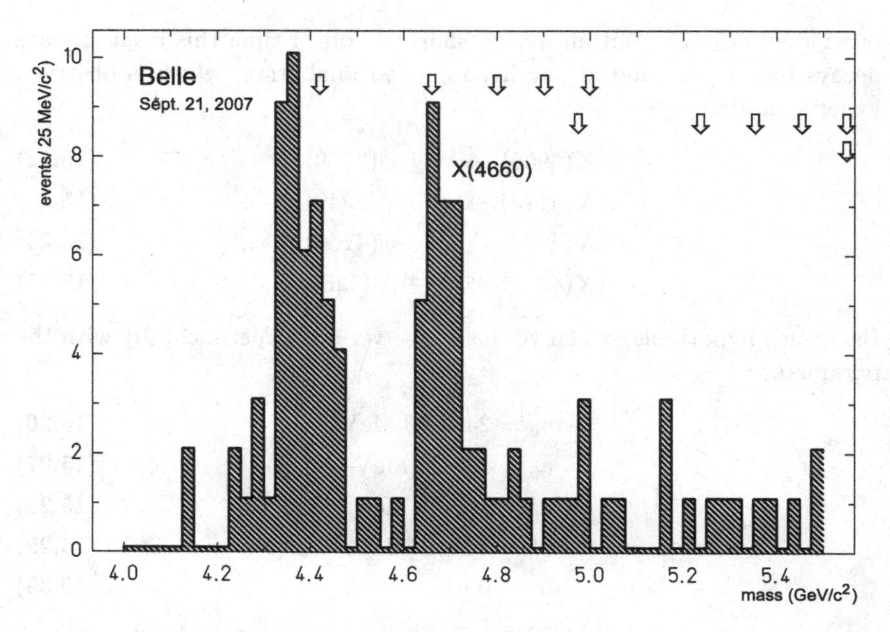


Fig. 15.3. Results from the Belle collaboration from September 2007 [Wang(2007)]. A new particle at $X(4660)$ is detected. Within the fractional level spectrum listed in Table 15.2 this particle was predicted as the $|41\rangle$-state of the fractional rotation group.

The current status of research leaves it merely as a speculation, whether 3 events are already a significant accumulation, to assert the existence of the $|50\rangle$-state to 4961[MeV] and of the $|60\rangle$-state to 5539[MeV]. Hence a new generation of experiments should be performed.

In this section we have collected a large number of arguments, which all demonstrate, that the charmonium spectrum may be understood consistently on the basis of a fractional rotation group with $\alpha \approx 2/3$.

Recently discovered particles $X(4260)$ and $X(4660)$ may be reproduced within the experimental errors.

There is no other model worldwide, which describes the properties of the charmonium spectrum with similar accuracy.

## 15.3 Phenomenology of meson spectra

The charmonium spectrum is just one example of all possible mesonic quark-antiquark states. In this section we will demonstrate, that a si-

Table 15.3. Optimum parameter sets $\{m_0, a_0, b_0\}$ and rms-error in percent for a fit with experimental meson spectra. Since the experimental situation is not sufficient to determine all parameters simultaneously, in the lower part of this table all predicted parameters are labelled with (*).

| meson | name | $\alpha$ | $m_0$ | $a_0$ | $b_0$ | $\Delta$[%] |
|---|---|---|---|---|---|---|
| $u\bar{u} \pm d\bar{d}$ | pions (light mesons) | 0.842 | 149.26 | 292.66 | 137.85 | 3.40 |
| $u\bar{s} \pm s\bar{u}$ | kaons | 0.814 | 489.73 | 281.56 | 399.15 | 0.48 |
| $s\bar{s}$ | strangeonium | 0.751 | 534.96 | 278.96 | 132.57 | 1.24 |
| $s\bar{c} \pm c\bar{s}$ | $D_s$-mesons | 0.716 | 1617.06 | 276.97 | 156.85 | 0.02 |
| $c\bar{c}$ | charmonium | 0.648 | 2461.63 | 453.60 | {107.85, 301.16 | 0.18 |
| $b\bar{b}$ | bottomonium | 0.736 | 8978.68 | 323.37 | 55.84 | 0.09 |
| $u\bar{c}, c\bar{u}$ | $D$-mesons | 0.75* | 1528.23 | 258.10 | 154.72 | 0.0* |
| $u\bar{b}, b\bar{u}$ | $B$-mesons | 0.80* | 4906.38 | 260.01 | 49.36 | 0.0* |
| $s\bar{b}, b\bar{s}$ | $B_s$-mesons | 0.75* | 4919.72 | 336.49 | 49.83 | 0.0* |
| $c\bar{b}, b\bar{c}$ | $B_c$-mesons | 0.71* | 5819.36 | 365* | 51* | 0.0* |

multaneous description of all mesonic ground state excitation spectra may be obtained with the mass formula (15.11) if we associate the quantum numbers $|LM\rangle$ with the corresponding mesonic excitation states.

In a first simplified approach we will treat the up- and down-quarks as similar and will investigate the ground state excitation spectra of mesons of the family $\{u, s, c, b\}$. In Figs. 15.4–15.8 these spectra are plotted.

The free parameters $\{m_0, a_0, b_0\}$ of the fractional model (15.11) were determined by a fit with the experimental data. For light mesons the experimental data base is broad enough. Consequently there are no problems using a fit procedure. In fact, the situation is much easier to handle than in the special case of charmonium, discussed in the previous chapter, since the determination of a single fractional magnetic field strength parameter $b_0$ suffices to describe the experimental data sufficiently.

For heavy quarks we have only a limited number of experimental data. Consequently it is essential to deduce systematic trends in order to specify appropriate parameter values. Using conventional phenomenological models a survey of strangeonium, charmonium and bottomonium states did not reveal a systematic trend for the parameter sets derived [Henriquez(1979)].

The fundamental new aspect for an interpretation of meson spectra within the framework of a fractional rotation group is the fact, that systematic trends for the parameters may be recognized and may be used for an universal description of all possible mesonic ground state excitations.

From Table 15.3 a systematic trend for the optimum fractional deriva-

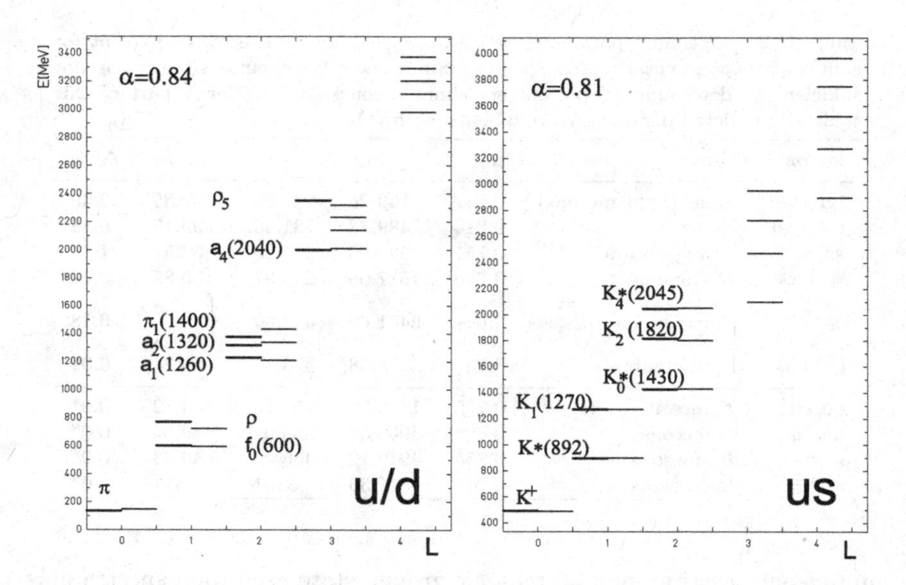


Fig. 15.4. Ground state meson excitation spectra for $u\bar{u}$ (light mesons) and $u\bar{s}$ (kaons). Thick lines represent experimental data [Nakamura(2010)]. Names and experimental masses are given. Thin lines denote the theoretical masses according to (15.11), values listed in Table 15.4.

tive parameter $\alpha$ for different mesonic systems may be deduced. These values are explained quantitatively, if we attribute to every quark and antiquark respectively a multiplicative quantum number $\alpha_q$, which we call fractional hypercharge in analogy to the standard hypercharge property associated with quarks:

$$\alpha_u = \sqrt{\tfrac{6}{7}} \quad \alpha_s = \sqrt{\tfrac{6}{8}} \quad \alpha_c = \sqrt{\tfrac{6}{9}} \quad \alpha_b = \sqrt{\tfrac{6}{8}} \tag{15.31}$$

Assigning these values to every single quark $\alpha$ follows from the product

$$\alpha = \alpha_{q_1 \bar{q}_2} = \alpha_{q_1} \alpha_{\bar{q}_2} \tag{15.32}$$

As a consequence, using (15.32) we predict $\alpha$-values for quark-antiquark combinations, where the experimental basis is yet too small to determine all parameters of the model (15.11) simultaneously. These predicted $\alpha$-values are labelled with (*) in Table 15.3.

The assumption of a fractional hypercharge reduces the number of free parameters of the model by one, which allows to calculate the spectra of $u\bar{c}$, $u\bar{b}$ and $s\bar{b}$ mesons on the basis of available experimental data. As a consequence, all meson ground state excitation spectra except $c\bar{b}$ may be

Table 15.4. complete listing of ground state meson excitations according to (15.11) with parameters from Table 15.3. For a given level in the first line identified particles and experimental masses in brackets, in the second line theoretical fit/prediction. All masses are given in MeV.

| level $\|LM\rangle$ | $u\bar{u}$ | $u\bar{s}$ | $s\bar{s}$ | $u\bar{c}$ | $s\bar{c}$ | $c\bar{c}$ | $u\bar{b}$ | $s\bar{b}$ | $c\bar{b}$ | $b\bar{b}$ |
|---|---|---|---|---|---|---|---|---|---|---|
| $\|00\rangle$ exp. | $\pi^0(134.976)$ | $K^+(493.667)$ | $\eta(547.51)$ | $f_2(1525)$ | $\eta_2(1617)$ | $D_2^*(2461.1)$ | | | $[B_2^*(5818)]$ | |
| $\|00\rangle$ th. | 149.26 | 489.73 | 534.96 | 1528.23 | 1617.06 | 2461.63 | 4906.38 | 4919.72 | 5819.36 | 8978.68 |
| $\|10\rangle$ exp. | $f_0(600)$ | $K^*(896)$ | $K(892)$ | $D(1864.5)$ | $D_s(1968.2)$ | $\eta_c(2980.4)$ | $B(5279.1)$ | $B_s(5367)$ | $B_c(6277)$ | $\eta_b(9388.90)$ |
| $\|10\rangle$ th. | 595.76 | 900.88 | 906.50 | 1856.51 | 1968.53 | 2989.58 | 5279.10 | 5367.00 | 6277.00 | 9400.27 |
| $\|11\rangle$ exp. | $\rho(775.5)$ | $K_1(1272)$ | $\phi(1019.46)$ | $D^*(2006.79)$ | $D_s^*(2112)$ | $J/\psi(3096.92)$ | $B^*(5325.1)$ | $B_s^*(5412.8)$ | | $\Upsilon(1s)(9460.3)$ |
| $\|11\rangle$ th. | 725.79 | 1274.15 | 1028.38 | 2011.92 | 2111.55 | 3086.63 | 5325.10 | 5412.80 | 6323.51 | 9451.42 |
| $\|20\rangle$ exp. | $a_1(1230)$ | $K_0^*(1425)$ | $f_1(1281.8)$ | | $D_{s0}^*(2317)$ | $\chi_{c0}(3414.76)$ | $B_2^*(5743)$ | $B_{sJ}^*(5853)$ | | $\chi_{b0}(9859.44)$ |
| $\|20\rangle$ th. | 1211.16 | 1428.80 | 1311.68 | 2213.22 | 2316.76 | 3419.65 | 5743.00 | 5853.00 | 6719.15 | 9841.41 |
| $\|21\rangle$ exp. | $a_2(1318.3)$ | $K_2(1816)$ | $f_1(1426.3)$ | – | $D_{s1}^*(2458.9)$ | $\chi_{c1}(3510.66)$ | | | | $\chi_{b1}(9892.78)$ |
| $\|21\rangle$ th. | 1341.19 | 1802.07 | 1433.56 | 2368.62 | 2459.78 | 3516.70 | 5789.00 | 5898.80 | 6765.56 | 9892.56 |
| $\|22\rangle$ exp. | $\pi_1(1376)$ | $K_4^*(2045)$ | $f_0(1507)$ | $D_2^*(2461.1)$ | $D_{s2}^*(2572.6)$ | $\chi_{c2}(3556.2)$ | $[B_2^*(5818)]$ | $B_{s2}^*$ | $B_{c2}^*$ | $\chi_{b2}(9912.1)$ |
| $\|22\rangle$ th. | 1434.13 | 2052.06 | 1503.73 | 2457.79 | 2535.35 | 3559.16 | 5818.94 | 5925.08 | 6789.42 | 9920.90 |
| $\|30\rangle$ exp. | $a_4(2001)$ | | $\phi_3(1854)$ | $D^*(2640)$ | $[D_{s1}^*(2690)]$ | $\psi(3772.92)$ | | $B_s^{**}$ | | $\Upsilon(3s)(10355.2)$ |
| $\|30\rangle$ th. | 2012.74 | 2099.90 | 1800.25 | 2642.85 | 2724.73 | 3763.12 | 6326.60 | 6438.37 | 7233.38 | 10366.7 |
| $\|31\rangle$ exp. | $\rho(2149)$ | | $f_2(1944)$ | | | $\psi(4039)$ | | | | |
| $\|31\rangle$ th. | 2142.77 | 2473.17 | 1922.13 | 2798.25 | 2867.75 | 4034.12 | 6372.60 | 6484.17 | 7279.79 | 10417.8 |
| $\|32\rangle$ exp. | | | $f_2(2011)$ | | | $\psi(4153)$ | | | | |
| $\|32\rangle$ th. | 2235.71 | 2723.16 | 1992.30 | 2887.42 | 2943.03 | 4152.68 | 6402.53 | 6510.45 | 7303.65 | 10446.2 |
| $\|33\rangle$ exp. | $\rho_5(2350)$ | | $f_4(2044)$ | | | $X(4259)$ | | | | |
| $\|33\rangle$ th. | 2321.99 | 2952.46 | 2055.05 | 2967.11 | 3009.45 | 4254.87 | 6429.85 | 6533.94 | 7324.63 | 10471.3 |
| $\|40\rangle$ exp. | | | $f_2(2339)$ | | | $\psi(4421)$ | | | | |
| $\|40\rangle$ th. | 2982.63 | 2898.27 | 2360.33 | 3134.97 | 3182.93 | 4413.10 | 7015.83 | 7108.88 | 7807.95 | 10963.5 |
| $\|41\rangle$ exp. | | | $f_6(2465)$ | | | $X(4660)$ | | | | $\Upsilon(11019)$ |
| $\|41\rangle$ th. | 3112.66 | 3271.54 | 2482.21 | 3290.37 | 3325.95 | 4684.09 | 7061.83 | 7154.68 | 7854.36 | 11014.6 |
| $\|42\rangle$ exp. | | | | | | $\psi(4790)(*)$ | | | | |
| $\|42\rangle$ th. | 3205.60 | 3521.53 | 2552.38 | 3379.54 | 3401.24 | 4802.66 | 7091.76 | 7180.96 | 7878.22 | 11042.9 |
| $\|43\rangle$ exp. | | | | | | | | | | |
| $\|43\rangle$ th. | 3291.88 | 3750.83 | 2615.13 | 3459.23 | 3467.66 | 4904.85 | 7119.08 | 7204.44 | 7899.20 | 11068.1 |
| $\|44\rangle$ exp. | | | | | | | | | | |
| $\|44\rangle$ th. | 3373.90 | 3966.94 | 2673.19 | 3532.94 | 3528.52 | 4996.81 | 7144.73 | 7226.17 | 7918.37 | 11091.3 |
| error % | 3.40 | 0.48 | 1.24 | *0.00 | 0.02 | 0.18 | *0.00 | *0.00 | *0.00 | 0.09 |

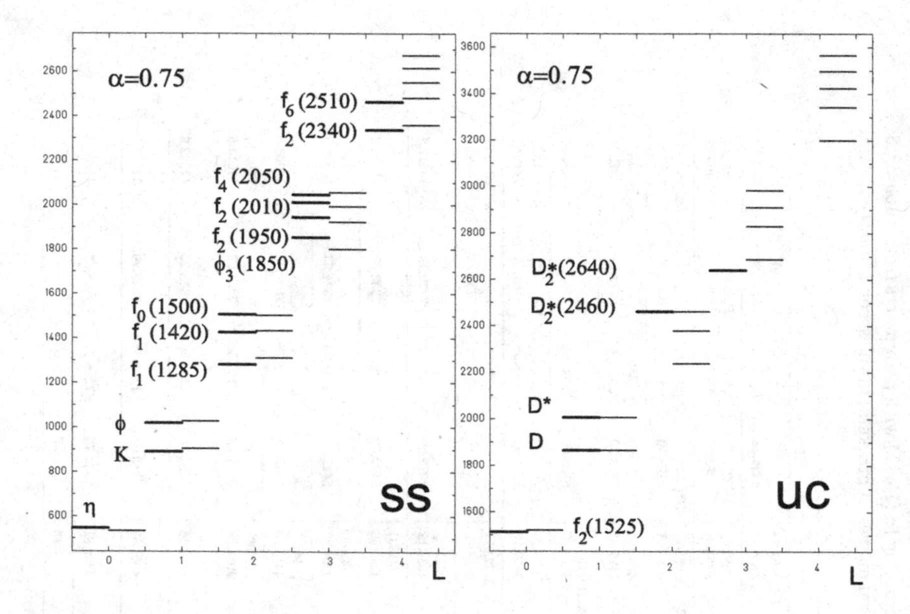


Fig. 15.5. Ground state meson excitation spectra for $s\bar{s}$ (strangeonium) and $u\bar{c}$ (D-mesons). Thick lines represent experimental data [Nakamura(2010)]. Names and experimental masses are given. Thin lines denote the theoretical masses according to (15.11), values listed in Table 15.4.

determined. Since for $c\bar{b}$-mesons until now only the $B_c$-meson has been verified experimentally, for this meson the number of model parameters has to be reduced to one.

A comparison of the optimum parameter sets for different quark-antiquark combinations reveals common aspects:

For all mesons except charmonium the parameter $a_0$ is of similar magnitude of order 280[MeV] with a weak tendency to higher values for heavy mesons.

In addition, the value for the fractional magnetic field strength $b_0$ for all mesons except kaons is about 150 [MeV]. Furthermore there seems to exist a saturation limit for the field strength $b_0 = B/m_q$ which is reached for charmonium. For bottomonium which is about three times heavier, the fitted $b_0$ turns out to be 55 [MeV], which is about 1/3 of the charmonium value. Most probably an additional enhancement of the field strength could exceed the critical energy density for the creation of a light quark-antiquark pair e.g. a pion.

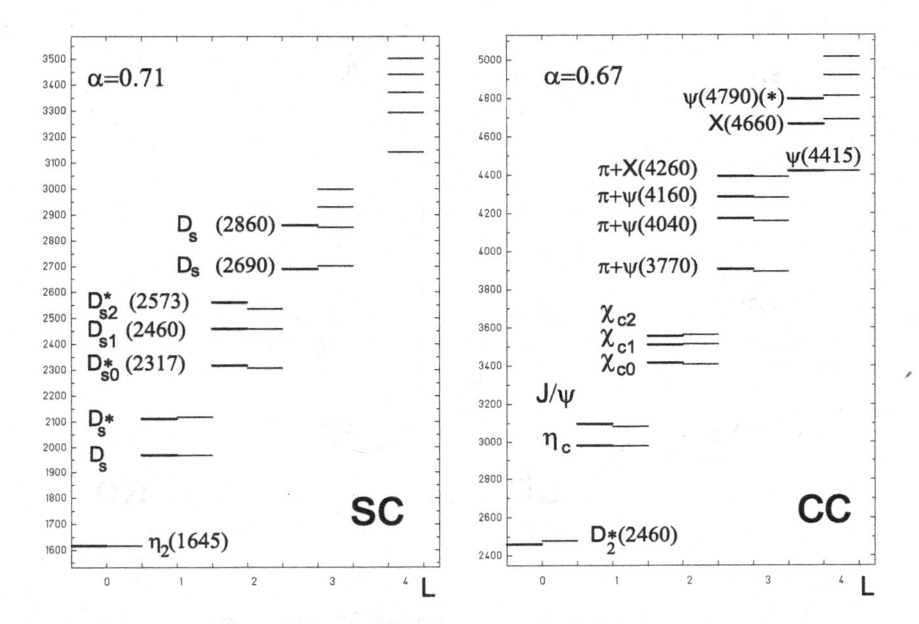


Fig. 15.6. Ground state meson excitation spectra for $s\bar{c}$ ($D_s$-mesons) and $c\bar{c}$ (charmonium). Thick lines represent experimental data [Nakamura(2010)]. Names and experimental masses are given. Thin lines denote the theoretical masses according to (15.11).

Consequently in view of a fractional mass formula (15.11) the different meson spectra have a great deal in common and systematic trends may be used to estimate the spectrum of $c\bar{b}$-mesons.

- From our considerations on a maximum magnetic field strength we deduce the following systematics for the magnetic field strength for the family of $Q_j\bar{b}$ mesons:
$$b_0 = c_0 m_b + c_1 m_{Q_j}, \qquad j = u, s, c, b \tag{15.33}$$
where $c_0, c_1$ are parameters and $m_{Q_j} = 4^j$ is the mass of the constituent quark (see discussion in the following section). It follows $b_0 = 51$ [MeV] for the $c\bar{b}$ meson.
- We assume a similar trend in the $a_0$ parameter for charmed and bottom mesons. We propose the relation:
$$\frac{a_0(c\bar{b})}{a_0(s\bar{b})} = \frac{a_0(s\bar{c})}{a_0(u\bar{c})} \tag{15.34}$$
With the values from Table 15.3 $a_0 = 365.00$ [MeV] follows for the $c\bar{b}$ meson.

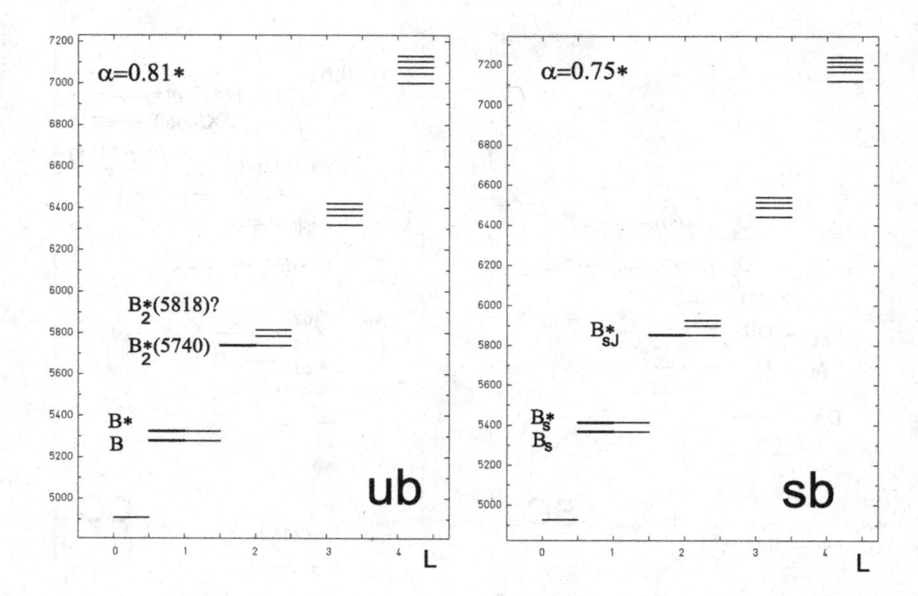


Fig. 15.7. Ground state meson excitation spectra for $u\bar{b}$ ($B$-mesons) and $s\bar{b}$ ($B_s$-mesons). Thick lines represent experimental data [Nakamura(2010)]. Names and experimental masses are given. Thin lines denote the theoretical masses according to (15.11).

With these assumptions the ground state spectra of all possible quark-antiquark combinations may be calculated. The resulting masses are presented in Table 15.4.

Based on the fractional mass formula (15.11) we found a general systematic in the ground state excitation spectra of mesons. Consequently we were able to classify all ground state excitation states for all possible mesons up to the bottom-quark as multiplets of the fractional rotation group $SO^\alpha(3)$.

The application of our fractional model reveals some unexpected characteristics: The predicted energies for the ground states $|00\rangle$ cannot be described with standard phenomenological potential models and are therefore an important indicator for the validity of the fractional concept.

Since we apply the Caputo fractional derivative, the ground state masses are directly determined by parameter $m_0$. We find a first family of well known mesons which we successfully associate with the calculated ground state masses.

- The ground state $|00\rangle$ of the $u\bar{u}$ system is predicted as 149[MeV]. Conse-

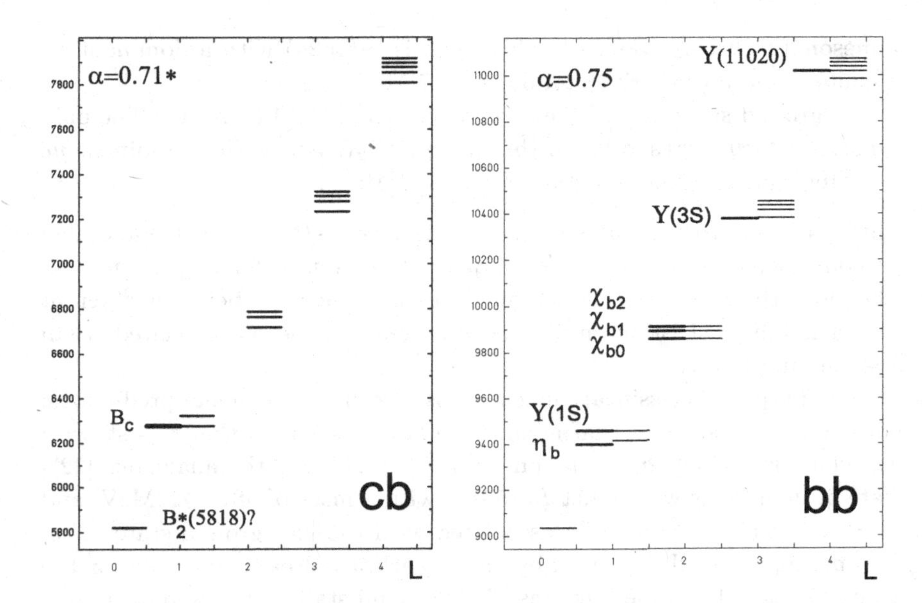


Fig. 15.8. Ground state meson excitation spectra for $c\bar{b}$ ($B_c$-mesons) and $b\bar{b}$ (bottomonium). Thick lines represent experimental data [Nakamura(2010)]. Names and experimental masses are given. Thin lines denote the theoretical masses according to (15.11).

quently we assign a pion with experimentally deduced mass of 135[MeV].

- The ground state $|00\rangle$ of the $u\bar{s}$ system is predicted as 489[MeV], which very well corresponds to the experimental mass of the $K^+$ kaon with 493[MeV].
- The ground state $|00\rangle$ of the $s\bar{s}$ is predicted as 534[MeV]. The lightest meson with a dominant $s\bar{s}$ content is the $\eta$-meson with an experimental mass of 547[MeV].

This family of ground states of light mesons is characterized by $J^P = 0^-$ and leaves the quark content of the meson unchanged.

For heavier mesons we propose experimental candidates for predicted ground states, which unexpectedly modify the quark contend of the meson-family considered:

- The ground state $|00\rangle$ of the $u\bar{c}$ is predicted as 1528.23[MeV]. The only meson which corresponds to this value is $f_2(1525)$ with a dominant $u\bar{u}$ content and an experimental mass 1525[MeV].
- The ground state $|00\rangle$ of the $s\bar{c}$ is predicted as 1617.06[MeV]. The only

meson which corresponds to this value is $\eta_2(1624)$ with a dominant $u\bar{s}$ content and an experimental mass 1617[MeV].

- The ground state $|00\rangle$ of the $c\bar{c}$ is predicted as 2461.63[MeV]. The only meson which corresponds to this value is $D_2^*(2460)$ with a dominant $u\bar{c}$ content and an experimental mass 2461.1[MeV].

This family of ground states is characterized by a $J^P = 2^+$ value and may be constructed by replacing the $\bar{c}$ quark by $\bar{u}$ quark for a given meson. Therefore the fractional model generates a connection between different meson families. Distinct meson spectra seem to be more related, than assumed until now.

The proposed classification scheme is also useful to make predictions: For the ground state of charmonium we have associated the $|22\rangle$- state of $u\bar{c}$, which is the $D_2^*(2640)$ meson with $J^p = 2^+$. For the analogous $|22\rangle$-state of $u\bar{b}$-mesons we predict $B_2^*(5818)$ with a mass of 5818.94 [MeV] and $J^p = 2^+$. This state may be associated with the $|00\rangle$ ground state of $c\bar{b}$-mesons, replacing the $\bar{c}$ quark by an $\bar{u}$ quark, as already observed for the ground states of charmed mesons. This ground state has a predicted mass of 5819.36 [MeV]. Within expected errors, this is an excellent agreement and pronounces the similarity of charm- and bottom-mesons in view of the proposed fractional mass formula.

Furthermore this is a first direct indication, that two different quark families $\{u, d, s\}$ and $\{c, b, t\}$ with sum of charges 0 and 1 respectively exist, which differ significantly in their $|00\rangle$ ground state properties. This may be interpreted as an indication, that in a fractional theory the quarks are grouped in triplets, while the standard model prefers generations of doublets ($\{u, d\}$, $\{s, c\}$ and $\{b, t\}$).

Summarizing our results we have demonstrated, that a fractional model may indeed explain the excitation spectra of all possible quark-antiquark combinations with an accuracy of better than 2%.

## 15.4 Metaphysics: About the internal structure of quarks

Finally we may speculate about an internal structure of quarks. We have calculated the ground state masses $m_0$ for all possible excitation spectra of mesons.

Let us assume that $m_0$ is the sum of the masses of two hypothetical constituent quarks ($Q_i$):

$$m_0(Q_1, Q_2) = m_{Q_1} + m_{Q_2} \tag{15.35}$$

We perform a fit procedure of parameter $m_0$ from Table 15.3 according to (15.35) and obtain:

$$Q_u \approx 75[\text{MeV}] \tag{15.36}$$

$$Q_s \approx 302[\text{MeV}] \tag{15.37}$$

$$Q_c \approx 1210[\text{MeV}] \tag{15.38}$$

$$Q_b \approx 4800[\text{MeV}] \tag{15.39}$$

A remarkable coincidence is the fact that $Q_s \approx 4Q_u$, $Q_c \approx 4Q_s$, $Q_b \approx 4Q_c$, the mass of every constituent quark is four times its predecessor. We obtain a rule

$$Q_j = m_u \, 4^j, \qquad j = 1, 2... \tag{15.40}$$

with $m_u = 18.92[\text{MeV}]$.

From this result we may speculate, that quarks exhibit an internal structure. Four constituents with mass $m_u$ constitute an up-constituent quark, 16 built up a strange-constituent quark *etc.* Therefore we naively guess, that this is a hint for an internal $SU(4)$-symmetry.

Furthermore we can make predictions for heavier quarkonia. We obtain

$$b^*\bar{b}^*(j = 5) \approx 38.7 \pm 5.0[\text{GeV}], \quad \alpha^* = \sqrt{\frac{6}{7}} \tag{15.41}$$

$$b^*\bar{b}^*(j = 6) \approx 155 \pm 20[\text{GeV}], \quad \alpha^* = 1 \tag{15.42}$$

So we are left with the problem, why these quarkonia have not been observed yet. It should be mentioned, that confinement is a natural property of free solutions of a fractional wave equation with $\alpha < 1$. Therefore the constituents of the $j = 6$ quarkonia are predicted to be quarks, which are not confined. As a consequence they are free particles. It is at least remarkable, that indeed there are direct observations of top-events near 171.3[GeV]. Therefore within the framework of fractional calculus, the top-quark could be interpreted as a $j = 6$ quarkonium state.

Summarizing our results, we found strong evidence, that the full excitation spectrum for baryons may successfully be interpreted as a fractional rotation spectrum based on the Riemann definition of the fractional derivative. Furthermore we have demonstrated, that the full variety of mesonic excitations may be interpreted as a fractional rotation spectrum too, if we apply the Caputo definition of the fractional derivative.

It is also remarkable, that the experimental hadron spectrum may be described quite successfully with the fractional mass formulas with $M \geq 0$.

We did not find any evidence for particles with $M < 0$. This is an indication, that chirality plays an important role in hadron physics.

The only difference between baryons and mesons within the framework of fractional calculus is the presence and absence of a zero-point energy contribution to the fractional rotational energy respectively, which we will interpret in the next chapter according to Racah's $SU(2)$ pairing model as a pairing gap.

This is a first indication, that Caputo- and Riemann fractional derivative describe distinct fundamental physical properties, which are directly related to fermionic and bosonic systems respectively.

Until now, we have investigated the properties of a single fractional particle. In the next section, we will give an additional surprising interpretation for a mixed Caputo-/Riemann- fractional derivative system with three independent particles.

# Chapter 16

# Higher Dimensional Fractional Rotation Groups

*We have demonstrated that the fractional extensions of the standard rotation group $SO(N)$ based on the Riemann and the Caputo fractional derivative definitions may successfully be used to describe excitation spectra of baryons and mesons respectively. Baryons are composite particle combinations of three quarks, while mesons are built up by a quark and an anti-quark. As a consequence, baryons are fermions and obey the Pauli-exclusion principle and Fermi–Dirac statistics, while mesons are bosons, which obey the Bose–Einstein statistics.*

*If we want to learn about the physical meaning of the difference between the Riemann- and Caputo based fractional derivative definition it will be helpful to investigate the properties of higher dimensional rotation groups.*

*Indeed we will find, that there are four different decompositions of the nine dimensional mixed fractional rotation group. For each decomposition we will determine the magic numbers associated with this symmetry.*

*As an astounding result we will realize, that two of these four sequences of magic numbers are realized in nature as magic nucleon numbers and as magic cluster numbers.*

*Consequently this is the first reasonable explanation for magic numbers based solely on symmetry considerations.*

## 16.1 The four decompositions of the mixed fractional $SO^\alpha(9)$

The nine dimensional Schrödinger equation

$$-\frac{\hbar^2}{2m}\sum_{i=1}^{9}\frac{\partial^2}{\partial x_i^2}\Psi(x_1,...,x_9) = E\Psi(x_1,...,x_9) \qquad (16.1)$$

may be considered as the quantum mechanical description of a single particle in a nine dimensional space. Alternatively we may interpret this equation as a description of $N = 3$ independent particles in three dimensional space:

$$-\frac{\hbar^2}{2m}\sum_{j=1}^{N=3}\sum_{i=1}^{3}\frac{\partial^2}{\partial x_{ij}^2}\Psi(x_{11},...,x_{33}) = E\Psi(x_{11},...,x_{33}) \tag{16.2}$$

In spherical coordinates this equation is separable. The angular part is a direct product of spherical harmonics $Y_{L_j M_j}$ which are eigenfunctions of the Casimir operators $C_j^2$ and $C_j^3$ of $SO(2)$ and $SO(3)$ respectively:

$$C_1^3(SO(3))|L_1M_1L_2M_2L_3M_3\rangle = L_1(L_1+1)|L_1M_1L_2M_2L_3M_3\rangle$$
$$L_1 = 0, 1, ... \tag{16.3}$$
$$C_1^2(SO(2))|L_1M_1L_2M_2L_3M_3\rangle = M_1|L_1M_1L_2M_2L_3M_3\rangle$$
$$M_1 = -L_1, ..., 0, ..., +L_1 \tag{16.4}$$
$$C_2^3(SO(3))|L_1M_1L_2M_2L_3M_3\rangle = L_2(L_2+1)|L_1M_1L_2M_2L_3M_3\rangle$$
$$L_2 = 0, 1, ... \tag{16.5}$$
$$C_2^2(SO(2))|L_1M_1L_2M_2L_3M_3\rangle = M_2|L_1M_1L_2M_2L_3M_3\rangle$$
$$M_2 = -L_2, ..., 0, ..., +L_2 \tag{16.6}$$
$$C_3^3(SO(3))|L_1M_1L_2M_2L_3M_3\rangle = L_3(L_3+1)|L_1M_1L_2M_2L_3M_3\rangle$$
$$L_3 = 0, 1, ... \tag{16.7}$$
$$C_3^2(SO(2))|L_1M_1L_2M_2L_3M_3\rangle = M_3|L_1M_1L_2M_2L_3M_3\rangle$$
$$M_3 = -L_3, ..., 0, ..., +L_3 \tag{16.8}$$

In terms of a group theoretical approach the nine dimensional rotation group $G$ may be decomposed into a chain of sub-algebras:

$$G \supset SO(3) \supset SO(3) \supset SO(3) \tag{16.9}$$

In the previous chapters we have discussed two different options for a possible extension to fractional rotation groups, namely

$$SO(n) \rightarrow \begin{cases} {}_{\mathrm{R}}SO^\alpha(n) & \text{Riemann} \\ {}_{\mathrm{C}}SO^\alpha(n) & \text{Caputo} \end{cases} \tag{16.10}$$

Consequently for a fractional extension of the nine dimensional rotation group $G^\alpha$ now 4 different decompositions exist with the following chain of sub-algebras:

$${}_{\mathrm{RRR}}G \supset {}_{\mathrm{R}}SO^\alpha(3) \supset {}_{\mathrm{R}}SO^\alpha(3) \supset {}_{\mathrm{R}}SO^\alpha(3) \tag{16.11}$$
$${}_{\mathrm{CRR}}G \supset {}_{\mathrm{C}}SO^\alpha(3) \supset {}_{\mathrm{R}}SO^\alpha(3) \supset {}_{\mathrm{R}}SO^\alpha(3) \tag{16.12}$$
$${}_{\mathrm{CCR}}G \supset {}_{\mathrm{C}}SO^\alpha(3) \supset {}_{\mathrm{C}}SO^\alpha(3) \supset {}_{\mathrm{R}}SO^\alpha(3) \tag{16.13}$$
$${}_{\mathrm{CCC}}G \supset {}_{\mathrm{C}}SO^\alpha(3) \supset {}_{\mathrm{C}}SO^\alpha(3) \supset {}_{\mathrm{C}}SO^\alpha(3) \tag{16.14}$$

Until now we have always assumed, that a fractional differential equation is based on a uniquely determined fractional derivative. Now we investigate rotation groups with mixed fractional derivative type.

We will demonstrate, that a new fundamental symmetry is established which will be used to determine the magic numbers in atomic nuclei and in electronic clusters accurately from a generalized point of view.

For that purpose, we will first introduce the necessary notation for a simultaneous treatment of mixed fractional rotation groups and will derive the corresponding level spectrum analytically.

## 16.2 Notation

We will investigate the spectrum of multi-dimensional fractional rotation groups for two different definitions of the fractional derivative, namely the Riemann- and Caputo fractional derivative. Both types are strongly related.

Starting with the definition of the fractional Riemann integral

$$ {}_{\mathrm{R}}I^{\alpha}\, f(x) = \begin{cases} ({}_{\mathrm{R}}I^{\alpha}_{+} f)(x) = \frac{1}{\Gamma(\alpha)} \int_0^x d\xi\, (x-\xi)^{\alpha-1} f(\xi) & x \geq 0 \\ ({}_{\mathrm{R}}I^{\alpha}_{-} f)(x) = \frac{1}{\Gamma(\alpha)} \int_x^0 d\xi\, (\xi-x)^{\alpha-1} f(\xi) & x < 0 \end{cases} \tag{16.15} $$

where $\Gamma(z)$ denotes the Euler $\Gamma$-function, the fractional Riemann derivative is defined as the result of a fractional integration followed by an ordinary differentiation:

$$ {}_{\mathrm{R}}\partial^{\alpha}_{x} = \frac{\partial}{\partial x}\, {}_{\mathrm{R}}I^{1-\alpha} \tag{16.16} $$

It is explicitly given by:

$$ {}_{\mathrm{R}}\partial^{\alpha}_{x}\, f(x) = \begin{cases} ({}_{\mathrm{R}}\partial^{\alpha}_{+} f)(x) = \frac{1}{\Gamma(1-\alpha)} \frac{\partial}{\partial x} \int_0^x d\xi\, (x-\xi)^{-\alpha} f(\xi) & x \geq 0 \\ ({}_{\mathrm{R}}\partial^{\alpha}_{-} f)(x) = \frac{1}{\Gamma(1-\alpha)} \frac{\partial}{\partial x} \int_x^0 d\xi\, (\xi-x)^{-\alpha} f(\xi) & x < 0 \end{cases} \tag{16.17} $$

The Caputo definition of a fractional derivative follows an inverted sequence of operations (16.16). An ordinary differentiation is followed by a fractional integration

$$ {}_{\mathrm{C}}\partial^{\alpha}_{x} = {}_{\mathrm{R}}I^{1-\alpha} \frac{\partial}{\partial x} \tag{16.18} $$

This results in:

$$ {}_{\mathrm{C}}\partial^{\alpha}_{x}\, f(x) = \begin{cases} ({}_{\mathrm{C}}\partial^{\alpha}_{+} f)(x) = \frac{1}{\Gamma(1-\alpha)} \int_0^x d\xi\, (x-\xi)^{-\alpha} \frac{\partial}{\partial \xi} f(\xi) & x \geq 0 \\ ({}_{\mathrm{C}}\partial^{\alpha}_{-} f)(x) = \frac{1}{\Gamma(1-\alpha)} \int_x^0 d\xi\, (\xi-x)^{-\alpha} \frac{\partial}{\partial \xi} f(\xi) & x < 0 \end{cases} \tag{16.19} $$

Applied to a function set $f(x) = x^{n\alpha}$ using the Riemann fractional derivative definition (16.17), we obtain:

$$ {}_{\mathrm{R}}\partial_x^\alpha \, x^{n\alpha} = \frac{\Gamma(1+n\alpha)}{\Gamma(1+(n-1)\alpha)} \, x^{(n-1)\alpha} \tag{16.20} $$

$$ = {}_{\mathrm{R}}[n] \, x^{(n-1)\alpha} \tag{16.21} $$

where we have introduced the abbreviation ${}_{\mathrm{R}}[n]$.

For the Caputo definition of the fractional derivative it follows for the same function set:

$$ {}_{\mathrm{C}}\partial_x^\alpha \, x^{n\alpha} = \begin{cases} \frac{\Gamma(1+n\alpha)}{\Gamma(1+(n-1)\alpha)} \, x^{(n-1)\alpha} & n > 0 \\ 0 & n = 0 \end{cases} $$

$$ = {}_{\mathrm{C}}[n] \, x^{(n-1)\alpha} \tag{16.22} $$

where we have introduced the abbreviation ${}_{\mathrm{C}}[n]$.

Both derivative definitions only differ in the case $n = 0$:

$$ {}_{\mathrm{C}}[n] = {}_{\mathrm{R}}[n] - \delta_{n0} \, {}_{\mathrm{R}}[0] \tag{16.23} $$

$$ = {}_{\mathrm{R}}[n] - \delta_{n0} \, \frac{1}{\Gamma(1-\alpha)} \tag{16.24} $$

where $\delta_{mn}$ denotes the Kronecker-$\delta$. We will rewrite equations (16.21) and (16.22) simultaneously, introducing the short hand notation

$$ {}_{\mathrm{R,C}}\partial_x^\alpha \, x^{n\alpha} = {}_{\mathrm{R,C}}[n] \, x^{(n-1)\alpha} \tag{16.25} $$

We now introduce the fractional angular momentum operators or generators of infinitesimal rotations in the $i, j$ plane on the $N$ dimensional Euclidean space:

$$ {}_{\mathrm{R,C}}L_{ij}(\alpha) = i\hbar(x_i^\alpha \, {}_{\mathrm{R,C}}\partial_j^\alpha - x_j^\alpha \, {}_{\mathrm{R,C}}\partial_i^\alpha) \tag{16.26} $$

which result from canonical quantization of the classical angular momentum definition. The commutation relations of the fractional angular momentum operators are isomorphic to the fractional extension of the rotational group $SO(N)$

$$ {}_{\mathrm{R,C}}[L_{ij}(\alpha), L_{kl}(\alpha)] = i\hbar_{\mathrm{R,C}} f_{ijkl}{}^{mn}(\alpha)_{\mathrm{R,C}} L_{mn}(\alpha) \tag{16.27} $$

$$ i, j, k, l, m, n = 1, 2, .., N $$

with structure coefficients ${}_{\mathrm{R,C}}f_{ijkl}{}^{mn}(\alpha)$. Their explicit form depends on the function set the fractional angular momentum operators act on and on the fractional derivative type used.

According to the group chain

$$_{\text{R,C}}SO^\alpha(3) \supset {}_{\text{R,C}}SO^\alpha(2) \tag{16.28}$$

there are two Casimir operators $\Lambda_i$, namely $\Lambda_2 = L_z(\alpha) = L_{12}(\alpha)$ and $\Lambda_3 = L^2(\alpha) = L_{12}^2(\alpha) + L_{13}^2(\alpha) + L_{23}^2(\alpha)$. We introduce the two quantum numbers $L$ and $M$, which completely determine the eigenfunctions $|LM\rangle$. It follows

$$_{\text{R,C}}\, L_z(\alpha)|LM\rangle = \hbar \,\text{sign}(M)\, _{\text{R,C}}\, [|M|]\, |LM\rangle \tag{16.29}$$

$$M = -L, -L+1, ..., \pm 0, ..., L$$

$$_{\text{R,C}}\, L^2(\alpha)|LM\rangle = \hbar^2\, _{\text{R,C}}\, [L]_{\text{R,C}}\, [L+1]\, |LM\rangle \tag{16.30}$$

$$L = 0, 1, 2, ...$$

where $|M|$ denotes the absolute value of $M$. In addition, on the set of eigenfunctions $|LM\rangle$, the parity operator $\Pi$ is diagonal and has the eigenvalues

$$\Pi|LM\rangle = (-1)^L|LM\rangle \tag{16.31}$$

Near $\alpha \approx 1$ there is a region of a rotational type of spectrum, while for $\alpha \approx 1/2$, the levels are nearly equidistant, which corresponds to a vibrational type of spectrum.

In addition, for decreasing $\alpha < 1$ higher angular momenta are lowered.

Only in the case $L = 0$ the spectra differ for the Riemann- and Caputo derivative. While for the Caputo derivative

$$_{\text{C}}L^2(\alpha)|00\rangle = 0 \tag{16.32}$$

because $_{\text{C}}[0] = 0$, using the Riemann derivative for $\alpha \neq 1$ there is a non vanishing contribution

$$_{\text{R}}L^2(\alpha)|00\rangle = \hbar^2\, _{\text{R}}[0]_{\text{R}}[1]|00\rangle = \hbar^2 \frac{\Gamma(1+\alpha)}{\Gamma(1-\alpha)}|00\rangle \tag{16.33}$$

In analogy to Racah's $SU(2)$ model of a pairing interaction [Racah(1943)] we may therefore interpret (16.33) as a pairing energy contribution, which in case of using the Caputo derivative definition for $L = 0$ leads to a lowering of the ground state energy.

In the next section we will demonstrate, that near the semiderivative $\alpha \approx 1/2$ we may interpret the $_{\text{R,C}}SO^\alpha(3)$ as a spherical representation of $U(1)$ and $U(1) \otimes U(1)_{\text{pairing}}$ respectively

$$_{\text{R}}SO^{\alpha=1/2}(3) \supset U(1) \tag{16.34}$$

$$_{\text{C}}SO^{\alpha=1/2}(3) \supset U(1) \otimes U(1)_{\text{pairing}} \tag{16.35}$$

if some specific symmetry requirements are imposed.

## 16.3 The nine dimensional fractional Caputo–Riemann–Riemann symmetric rotor

We use group theoretical methods to construct higher dimensional representations of the fractional rotation groups ${}_{\mathrm{R,C}}SO^{\alpha}(3)$.

As an example of physical relevance we will investigate the properties of the nine dimensional fractional rotation group ${}_{\mathrm{CRR}}G$ (16.12) with the following chain of sub algebras:

$$ {}_{\mathrm{CRR}}G \supset {}_{\mathrm{C}}SO^{\alpha}(3) \supset {}_{\mathrm{R}}SO^{\alpha}(3) \supset {}_{\mathrm{R}}SO^{\alpha}(3) \tag{16.36} $$

We associate a Hamiltonian $H$, which can now be written in terms of the Casimir operators of the algebras appearing in the chain and can be analytically diagonalized in the corresponding basis. The Hamiltonian is explicitly given as:

$$ H = \frac{\omega_1}{\hbar}\,{}_{\mathrm{C}}L_1^2(\alpha) + \frac{\omega_2}{\hbar}\,{}_{\mathrm{R}}L_2^2(\alpha) + \frac{\omega_3}{\hbar}\,{}_{\mathrm{R}}L_3^2(\alpha) \tag{16.37} $$

with the free parameters $\omega_1, \omega_2, \omega_3$ and the basis is $|L_1M_1L_2M_2L_3M_3\rangle$. Furthermore, we demand the following symmetries:

First, the wave functions should be invariant under parity transformations, which according to (16.31) leads to the conditions

$$ L_1 = 2n_1, \quad L_2 = 2n_2, \quad L_3 = 2n_3, \quad n_1, n_2, n_3 = 0, 1, 2, 3, ... \tag{16.38} $$

second, we require

$$ {}_{\mathrm{C}}L_{z_1}(\alpha)|L_1M_1L_2M_2L_3M_3\rangle = +\hbar_{\mathrm{C}}[L_1]|L_1M_1L_2M_2L_3M_3\rangle \tag{16.39} $$

$$ {}_{\mathrm{R}}L_{z_2}(\alpha)|L_1M_1L_2M_2L_3M_3\rangle = +\hbar_{\mathrm{R}}[L_2]|L_1M_1L_2M_2L_3M_3\rangle \tag{16.40} $$

$$ {}_{\mathrm{R}}L_{z_3}(\alpha)|L_1M_1L_2M_2L_3M_3\rangle = +\hbar_{\mathrm{R}}[L_3]|L_1M_1L_2M_2L_3M_3\rangle \tag{16.41} $$

which leads to the conditions

$$ M_1 = 2n_1, \quad M_2 = 2n_2, \quad M_3 = 2n_3, \quad n_1, n_2, n_3 = 0, 1, 2, 3, ... \tag{16.42} $$

and reduces the multiplicity of a given $|2n_1M_12n_2M_22n_3M_3\rangle$ set to 1.

With these conditions, the eigenvalues of the Hamiltonian (16.37) are given as

$$ E(\alpha) = \hbar\omega_1\,{}_{\mathrm{C}}[2n_1]_{\mathrm{C}}[2n_1+1] + \hbar\omega_2\,{}_{\mathrm{R}}[2n_2]_{\mathrm{R}}[2n_2+1] + \hbar\omega_3\,{}_{\mathrm{R}}[2n_3]_{\mathrm{R}}[2n_3+1] \tag{16.43} $$

$$ [9pt] = \sum_{i=1}^{3} \hbar\omega_i \frac{\Gamma(1+(2n_i+1)\alpha)}{\Gamma(1+(2n_i-1)\alpha)} - \delta_{n_1 0}\hbar\omega_1 \frac{\Gamma(1+\alpha)}{\Gamma(1-\alpha)} \tag{16.44} $$

$$ n_1, n_2, n_3 = 0, 1, 2, $$

on a basis $|2n_12n_12n_22n_22n_32n_3\rangle$.

This is the major result of our derivation. We call this model the Caputo–Riemann–Riemann symmetric rotor.

In the next section we will investigate the properties of this model near the semiderivative $\alpha = 1/2$ and will present a surprising coincidence with the magic numbers of nuclei.

## 16.4 Magic numbers of nuclei

The experimental evidence for discontinuities in the sequence of atomic masses, $\alpha$- and $\beta$-decay systematic and binding energies of nuclei suggests the existence of a set of magic proton and neutron numbers, which can be described successfully by single particle shell models with a heuristic spin-orbit term [Goeppert-Mayer(1949), Haxel(1949)]. The most prominent representative is the phenomenological Nilsson model [Nilsson(1955)] with an anisotropic oscillator potential:

$$V(x_i) = \sum_{i=1}^{3} \frac{1}{2} m\omega_i^2 x_i^2 - \hbar\omega_0 \kappa(2\vec{l}\vec{s} + \mu l^2) \tag{16.45}$$

Although these models are flexible enough to reproduce the experimental results, they lack a deeper theoretical justification, which becomes obvious, when extrapolating the parameters $\kappa$, $\mu$, which determine the strength of the spin orbit and $l^2$ term to the region of super-heavy elements [Hofmann(2000)].

Hence it seems tempting to describe the experimental data with alternative methods. Typical examples are microscopic Hartree–Fock calculations with a uniform nucleon-nucleon interaction of e.g. Skyrme type [Vautherin(1972)] or relativistic mean field theories [Rufa(1988), Bender(2001)], where nucleons are described by the Dirac-equation and the interaction is mediated by mesons. Although a spin orbit force is unnecessary in these models, different parameterizations predict different shell closures [Rutz(1997), Kruppa(2000)].

Therefore the problem of a theoretical foundation of magic numbers remains an open question since Elsasser [Elsasser(1933)] raised the problem 77 years ago.

We will demonstrate, applying the Caputo–Riemann–Riemann symmetric rotor (16.43), that a new fundamental dynamic symmetry is established, which determines the magic numbers for protons and neutrons and furthermore describes the ground state properties like binding energies

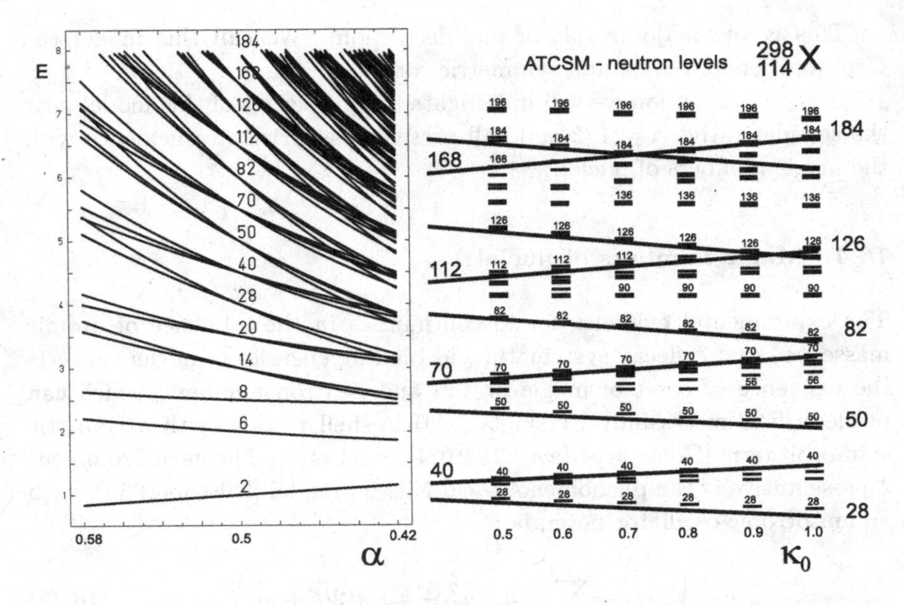


Fig. 16.1. On the left the energy spectrum $E(\alpha)$ from (16.44) for the spherical case (16.46) in units of $\hbar\omega_0$ for the Caputo–Riemann–Riemann symmetric rotor near the ideal vibrational case $\alpha = 1/2$ is presented. The right diagram shows the neutron energy levels for the spherical nucleus $^{298}_{114}X$ calculated within the framework of the asymmetric two centre shell model (ATCSM) [Maruhn(1972)], which exactly corresponds to the Nilsson shell model (16.45) near the ground state as a function of increasing strength of the spin-orbit term ($\kappa_0 \kappa l \vec{s}$) increasing from 50% to 100% of the recommended $\kappa$ value, while the $\mu l^2$ value is kept constant. The transition from magic numbers of the standard three dimensional harmonic oscillator levels (16.49) to the shifted set of magic numbers (16.52) is pointed out with thick lines. Left and right figure therefore show a similar behaviour for the energy levels.

and ground state quadrupole deformations of nuclei with reasonable accuracy.

On the left of Fig. 16.1 we have plotted the energy levels of the Caputo–Riemann–Riemann fractional symmetric rotor (16.43) in the vicinity of $\alpha \approx 1/2$ for the case

$$\omega_1 = \omega_2 = \omega_3 = \omega_0 \tag{16.46}$$

which we denote as the spherical case.

For the idealised case $\alpha = 1/2$, using the relation $\Gamma(1+z) = z\Gamma(z)$ the level spectrum (16.43) is simply given by:

$$E(\alpha = 1/2) = \hbar\omega_0 \left( n_1 + n_2 + n_3 + \frac{3}{2} - \frac{1}{2}\delta_{n_1 0} \right) \tag{16.47}$$

According (16.34) and (16.35) near $\alpha = 1/2$ the quantum numbers $n_i$ may be interpreted as eigenvalues of the $U(1)$ number operator. For $n_1 \neq 0$ the energy spectrum (16.47) is the well known spectrum of the three dimensional harmonic oscillator. We introduce the quantum number $N$ as

$$N = n_1 + n_2 + n_3 \tag{16.48}$$

Assuming a two-fold spin degeneracy of the energy levels, we obtain a first set $n_{\text{magic }1}$ of magic numbers $n_{\text{magic}}$

$$\begin{aligned} n_{\text{magic }1} &= \frac{1}{3}(N+1)(N+2)(N+3), \qquad N = 1, 2, 3, ... & (16.49)\\ &= 8, 20, 40, 70, 112, 168, 240, ... & (16.50)\end{aligned}$$

which correspond to the standard three dimensional harmonic oscillator at energies

$$E = \hbar\omega_0(N + 3/2) \tag{16.51}$$

In addition, for $n_1 = 0$, which corresponds to the $|00\,2n_2 2n_2\,2n_3 2n_3\rangle$ states with a $2\sum_{n=0}^{N} n = N(N+1)$-fold multiplicity, we obtain a second set $n_{\text{magic }2}$ of magic numbers

$$\begin{aligned} n_{\text{magic }2} &= n_{\text{magic }1} - N(N+1), \qquad N = 0, 1, 2, 3, ... & (16.52)\\ &= \frac{1}{3}(N+1)((N+2)(N+3) - 3N) & (16.53)\\ &= \frac{1}{3}(N+1)((N+1)^2 + 5) & (16.54)\\ &= 2, 6, 14, 28, 50, 82, 126, 184, 258, ... & (16.55)\end{aligned}$$

at energies

$$E = \hbar\omega_0(N + 1) \tag{16.56}$$

which is shifted by the amount $-\frac{1}{2}\hbar\omega_0$ compared to the standard three dimensional harmonic oscillator values.

From Fig. 16.1 it follows, that for $\alpha < 1/2$ the second set $n_{\text{magic }2}$ of energy levels falls off more rapidly than the levels of set $n_{\text{magic }1}$. As a consequence for decreasing $\alpha$ the magic numbers $n_{\text{magic }1}$ die out successively. On the other hand, for $\alpha > 1/2$ the same effect causes the magic numbers $n_{\text{magic }1}$ to survive.

We want to emphasize, that the described behaviour for the energy levels in the region $\alpha < 1/2$ may be directly compared to the influence of a $ls$-term in phenomenological shell models. As an example, on the right hand side of Fig. 16.1 a sequence of neutron levels for the super-heavy element

$^{298}_{114}X$ calculated with the asymmetric two centre shell model (ATCSM) [Maruhn(1972)], which exactly corresponds to the Nilsson shell model near the spherical ground state, with increasing strength of the $ls$-term from 50% to 100% is plotted. It shows, that the $n = 168$ gap breaks down at about 70% and the $n = 112$ gap at about 90% of the recommended $\kappa$-value for the $ls$-term. This corresponds to an $\alpha \approx 0.46$ value, since in the Caputo–Riemann–Riemann symmetric rotor the $n = 168$ gap breaks down at $\alpha = 0.466$, the $n = 112$ gap at $\alpha = 0.460$ and the $n = 70$ gap vanishes at $\alpha = 0.453$.

We conclude, that the Caputo–Riemann–Riemann symmetric rotor predicts a well defined set of magic numbers. This set is a direct consequence of the underlying dynamic symmetries of the 3 fractional rotation groups involved. It is indeed remarkable, that the same set of magic numbers is realized in nature as magic proton and neutron numbers.

In the next section we will demonstrate, that the proposed analytical model is an appropriate tool to describe the ground state properties of nuclei.

## 16.5 Ground state properties of nuclei

We will use the Caputo–Riemann–Riemann symmetric rotor (16.43) as a dynamic shell model for a description of the microscopic part of the total energy $E_{\text{tot}}$ of the nucleus.

$$E_{\text{tot}} = E_{\text{macroscopic}} + E_{\text{microscopic}} \tag{16.57}$$

$$= E_{\text{macroscopic}} + \delta U + \delta P \tag{16.58}$$

where $\delta U$ and $\delta P$ denote the shell- and pairing energy contributions.

For the macroscopic contribution we use the finite range liquid drop model (FRLDM) proposed by Möller [Möller(1995)] using the original parameters, except the value for the constant energy contribution $a_0$, which will be used as a free parameter for a fit with the experimental data.

As the primary deformation parameter we use the ellipsoidal deformation $Q$:

$$Q = \frac{b}{a} = \frac{\omega_3}{\omega_1} = \frac{\omega_3}{\omega_2} \tag{16.59}$$

where $a, b$ are the semiaxes of a rotational symmetric ellipsoid. Consequently a value $Q < 1$ describes prolate and a value of $Q > 1$ describes

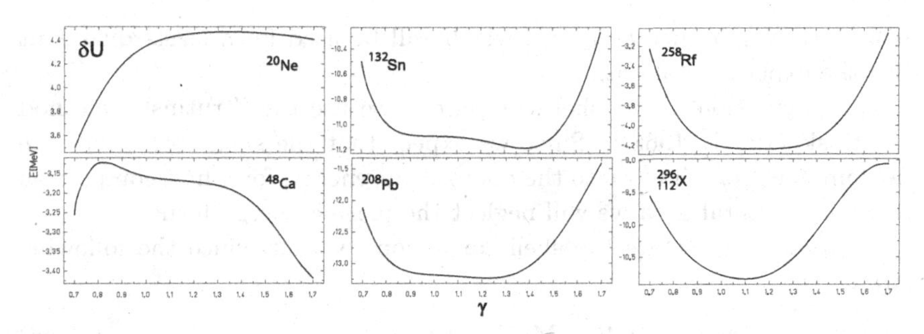


**Fig. 16.2.** As a test of the plateau condition $\partial U/\partial\gamma = 0$ for the Strutinsky shell correction method, the total shell correction energy $\delta U = \delta U_P + \delta U_N$ is plotted as a function of $\gamma$ for different nuclei.

oblate shapes. In order to relate the ellipsoidal deformation $Q$ to the quadrupole deformation $\epsilon_2$ used by Möller, we use the relation:

$$Q = 1 - 1.43085\epsilon_2 + 0.707669\epsilon_2^2 \tag{16.60}$$

which is a result from a least-squares fit and quadratic approximation of equipotential surfaces.

Furthermore we extend the original FRLDM-model introducing an additional curvature energy term $V_R(Q)$, which describes the interaction of the nucleus with the collective curved coordinate space [Herrmann(2008a)]:

$$V_R(Q) = -a_R B_R A^{-5/3} \tag{16.61}$$

where $A$ is the nucleon number, $a_R$ is the curvature parameter given in [MeV] and the relative curvature energy $B_R(Q)$ given as:

$$B_R(Q) = 9\,Q^{16/3}\left(\frac{199 - 288\ln(2)}{(2+Q^2)(266 - 67Q^2 + 96(Q^2-4)\ln(2))}\right)^2 \tag{16.62}$$

which is normalized relative to a sphere $B_R(Q=1) = 1$.

Therefore the total energy may be split into

$$E_{\text{tot}} = E_{\text{mac}} + E_{\text{mic}} \tag{16.63}$$

where

$$E_{\text{mac}}(a_0, a_R) = \text{FRLDM}(a_0, Q=1) + V_R(Q=1, a_R) \tag{16.64}$$

$$\begin{aligned} E_{\text{mic}}(a_0, a_R, Q) = &+\delta U + \delta P + \text{FRLDM}(a_0, Q) + V_R(Q, a_R) \\ &-(\text{FRLDM}(a_0, Q=1) + V_R(Q=1, a_R)) \end{aligned} \tag{16.65}$$

with two free parameters $a_0, a_R$, which will be used for a least-squares fit with the experimental data.

For calculation of the shell corrections we use the Strutinsky method [Strutinsky(1967b, 1968)]. Since we expect that the shell corrections are the dominant contribution to the microscopic energy, for a first comparison with experimental data we will neglect the pairing energy term.

In order to calculate the shell corrections, we introduce the following parameters:

$$\hbar\omega_0 = 38A^{-\frac{1}{3}}[MeV] \tag{16.66}$$

$$\omega_1 = \omega_0 Q^{-\frac{1}{3}} \tag{16.67}$$

$$\omega_2 = \omega_0 Q^{-\frac{1}{3}} \tag{16.68}$$

$$\omega_3 = \omega_0 Q^{\frac{2}{3}} \tag{16.69}$$

$$\alpha_Z = \begin{cases} 0.46 + 0.000220\, Z & Z > 50 \\ 0.2469 + 0.00448\, Z & 28 < Z \leq 50 \\ 0.2793 + 0.00332\, Z & Z \leq 28 \end{cases} \tag{16.70}$$

$$\alpha_N = \begin{cases} 0.41 + 0.000200\, N & N > 50 \\ 0.3118 + 0.00216\, N & 28 < N \leq 50 \\ 0.2793 + 0.00332\, N & N \leq 28 \end{cases} \tag{16.71}$$

$$\gamma = 1.1\,\hbar\omega_0 \tag{16.72}$$

$$m = 4 \tag{16.73}$$

$$a_0 = 2.409[\text{MeV}] \tag{16.74}$$

$$a_R = 15.0[\text{MeV}] \tag{16.75}$$

Input parameters are the number of protons $Z$, number of neutrons $N$, the nucleon number $A = N + Z$, and the ground state quadrupole deformation $\epsilon_2$.

The values obtained include

- the frequencies (16.67)–(16.69), which are related to the quadrupole deformation $\epsilon_2$ via (16.60),
- the fractional derivative coefficients for protons (16.70) and neutrons (16.71) which determine the level spectrum for protons and neutrons for the proton and neutron part of the shell correction energy respectively from a fit of the set of nuclides ${}^{56}_{28}$Ni, ${}^{100}_{50}$Sn, ${}^{132}_{50}$Sn, ${}^{208}_{82}$Pb and from the requirement, that the neutron shell correction for ${}^{100}_{50}$Sn should amount about $-5.1[MeV]$,

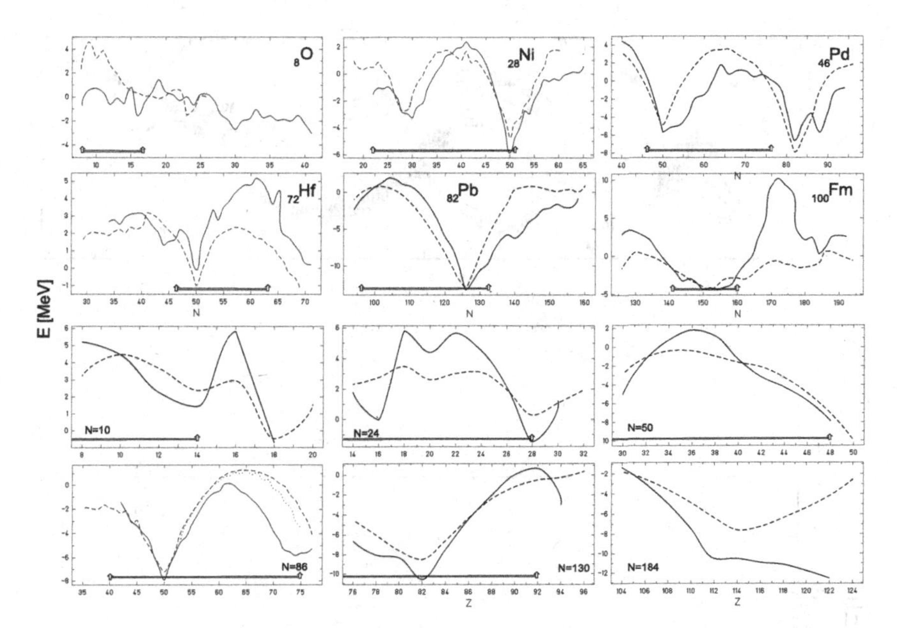


Fig. 16.3. Comparison of calculated shell corrections $\delta U$ from the Caputo–Riemann–Riemann symmetric rotor (16.44) with the parameter set (16.66)–(16.73) (thick line) with the tabulated $E_{\text{mic}}$ from Möller [Möller(1995)] (dashed line). Upper two rows show values for a given $Z$ as a function of $N$, lower two rows for a given $N$ as a function of $Z$. Bars indicate the experimentally known region. The original $\epsilon_2$ values from [Möller(1995)] are used, which is the main source of error.

- (16.72) from the plateau condition $\partial U/\partial \gamma = 0$ (see Fig. 16.2) and
- (16.73) the order of included Hermite polynomials for the Strutinsky shell correction method. Finally
- $\hbar\omega_0, a_0, a_R$ from a fit of the experimental mass excess given in [Audi(2003)].

We compare our results for the microscopic energy contribution $E_{\text{mic}}$ with data from Möller et al. [Möller(1995)] and use their tabulated $\epsilon_2$ values. They have not only listed data for experimental masses but also predictions for regions, not yet confirmed by experiment.

In Fig. 16.3 we compare the calculated $\delta U$ values with the tabulated $E_{\text{mic}}$, which is justified for almost spherical shapes ($\epsilon_2 \approx 0$). The results agree very well within the expected errors (which are estimated $\approx 2\,[MeV]$ for the pairing energy and $0.5\,[MeV]$ for $E_{\text{mic}}$), especially in the region of experimentally known nuclei.

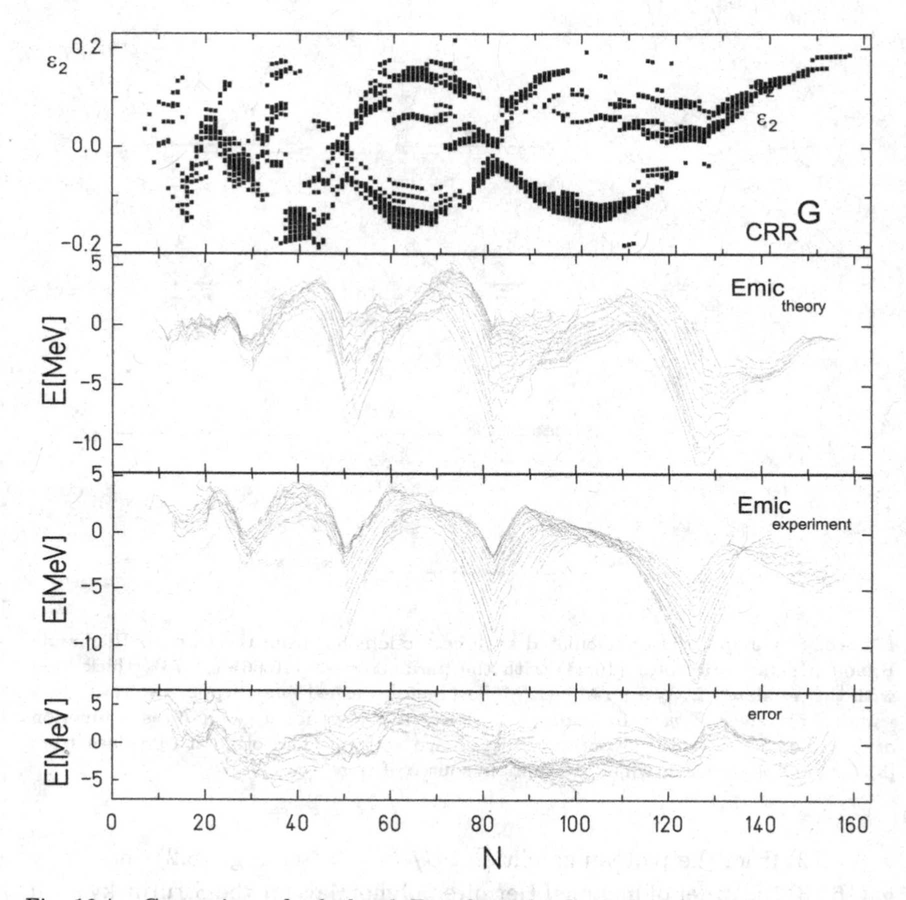


Fig. 16.4. Comparison of calculated $E_{\text{mic}}$ from the Caputo–Riemann–Riemann symmetric rotor (16.44) with the parameter set (16.66)–(16.75), minimized with respect to $\epsilon_2$ with the experimental masses from Audi [Audi(2003)] as a function of $N$. From top to bottom the minimized $\epsilon_2$ values, theoretical $E_{\text{mic}}$, experimental microscopic contribution from the difference of experimental mass excess and macroscopic FRLDM energy and error in $[MeV]$ are plotted.

A remarkable difference between the calculated shell correction and tabulated $E_{\text{mic}}$ from Möller occurs for spherical super-heavy elements ($N = 184$, last picture in Fig. 16.3).

While phenomenological shell models predict a pronounced minimum in the shell correction energy for $Z = 114$ [Myers(1966), Mosel(1969), Niyti(2010)] the situation is quite different for the rotor model, where two magic shell closures at $Z = 112$ and $Z = 126$ are given, but the $Z = 112$ shell closure is not strong enough to produce a local minimum in the shell correction energy plot as a function of $Z$. Instead, between $Z = 112$ and $Z = 126$, a slightly falling energy plateau emerges, which makes the full region promising candidates for stable, long-lived super-heavy elements.

While this result contradicts predictions made with phenomenological shell models, it supports recent results obtained with relativistic mean field models [Bender(2001)], which predict a similar behaviour in the region of super-heavy elements as the proposed rotor model.

In Fig. 16.4 we have covered the complete region of available experimental data for nuclides and compare the calculated theoretical microscopic energy contribution minimized with respect to the deformation with the experimental values. The influence of shell closures is very clear. The rms-error is about 2.4[MeV]. The maximum deviation occurs between closed magic shells.

Therefore in the next section we will introduce a generalization of the proposed fractional rotor model, which not only determines the magic numbers accurately but in addition determines the fine structure of the single particle spectrum correctly.

## 16.6 Fine structure of the single particle spectrum: the extended Caputo–Riemann–Riemann symmetric rotor

In the previous section we have demonstrated, that the Caputo–Riemann–Riemann symmetric rotor correctly determines the magic numbers in the single particle spectra for neutrons and protons. However, there remains a significant difference between calculated and experimental ground state masses for nuclei with nucleon numbers far from magic shell closures. This indicates that the fine structure of the single particle levels is not yet correctly reproduced.

We therefore propose the following generalization of the Caputo–Riemann–Riemann symmetric rotor group:

$$_{\mathrm{C3C2R3R3}}G \supset {}_{\mathrm{C}}SO^{\alpha}(3) \supset {}_{\mathrm{C}}SO^{\alpha}(2) \supset {}_{\mathrm{R}}SO^{\alpha}(3) \supset {}_{\mathrm{R}}SO^{\alpha}(3) \qquad (16.76)$$

with the Casimir operators (16.29) and (16.30) it follows for the Hamiltonian $H$:

$$H = \frac{\omega_1}{\hbar} {}_{\mathrm{C}}L_1^2(\alpha) + B\omega_{0\mathrm{C}}L_{z_1}(\alpha) + \frac{\omega_2}{\hbar} {}_{\mathrm{R}}L_2^2(\alpha) + \frac{\omega_3}{\hbar} {}_{\mathrm{R}}L_3^2(\alpha) \qquad (16.77)$$

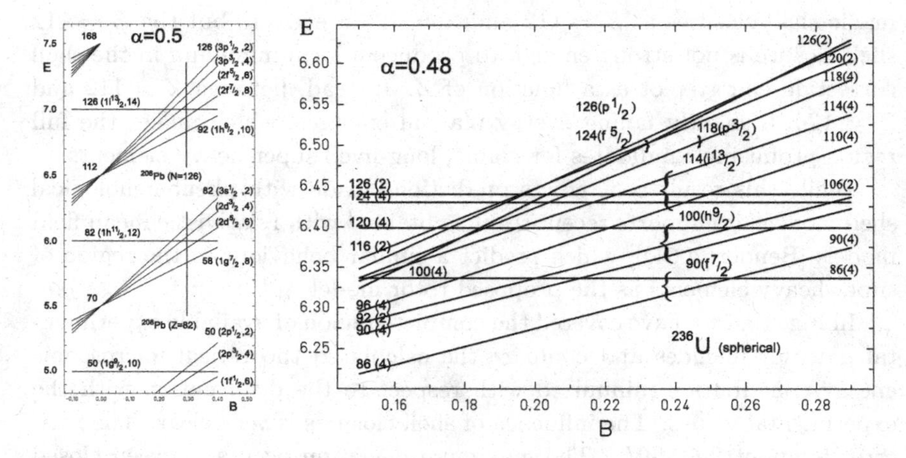


Fig. 16.5. For $\alpha = 1/2$, on the left side the level spectrum for the extended Caputo–Riemann–Riemann symmetric rotor (16.79) is plotted as a function of the fractional magnetic field strength $B$. The levels are labelled according to the corresponding $[Nlj_z]$ Nilsson scheme and the multiplicity is given. The level sequence reproducing the experimental data for lead is given for $N = 126$ and $Z = 82$, see also [Rufa(1988)]. For $\alpha = 0.48$ the resulting level sequence near $N \approx 126$ is plotted on the right. At $B \approx 0.25$ the resulting spectrum coincides with the corresponding spherical Nilsson level spectrum. Brackets indicate the proposed appropriate combinations of rotor levels.

with the free parameters $\omega_1, \omega_2, \omega_3, B$, where $B$ may be called fractional magnetic field strength in units $[\hbar\omega_0]$, since this Hamiltonian is the extension of the fractional Zeeman effect to nine dimensional space.

Demanding the same symmetries (16.38), (16.39) as in the case of the symmetric Caputo–Riemann–Riemann rotor, the eigenvalues of the Hamiltonian (16.77) are given as

$$\begin{aligned} E(\alpha) &= \hbar\omega_1\, {}_{\mathrm{C}}[2n_1]_{\mathrm{C}}[2n_1+1] + B\hbar\omega_0\, {}_{\mathrm{C}}[2n_1] \\ &\quad + \hbar\omega_2\, {}_{\mathrm{R}}[2n_2]_{\mathrm{R}}[2n_2+1] + \hbar\omega_3\, {}_{\mathrm{R}}[2n_3]_{\mathrm{R}}[2n_3+1] \qquad (16.78) \\ &= \sum_{i=1}^{3} \hbar\omega_i \frac{\Gamma(1+(2n_i+1)\alpha)}{\Gamma(1+(2n_i-1)\alpha)} - \delta_{n_1 0}\hbar\omega_1 \frac{\Gamma(1+\alpha)}{\Gamma(1-\alpha)} \\ &\quad + B\hbar\omega_0 \frac{\Gamma(1+2n_1\alpha)}{\Gamma(1+(2n_1-1)\alpha)} - \delta_{n_1 0} B\hbar\omega_0 \frac{1}{\Gamma(1-\alpha)} \\ &\qquad n_1, n_2, n_3 = 0, 1, 2, \qquad (16.79) \end{aligned}$$

on a basis $|2n_1 2n_1 2n_2 2n_2 2n_3 2n_3\rangle$.

We call this model the extended Caputo–Riemann–Riemann symmetric rotor. The additional ${}_{\mathrm{C}}L_{z_1}(\alpha)$ term yields a level splitting of the harmonic oscillator set of magic numbers $n_{\text{magic }1}$ (16.49), while the multiplicity of the $n_{\text{magic }2}$ set (16.52) remains unchanged, since this set is characterized by $n_1 = 0$. This is exactly the behaviour needed to describe the experimentally observed fine structure, as can be deduced from the right hand side of Fig. 16.1.

In order to clearly demonstrate the influence of the additional term, we first investigate the level spectrum for the spherical (16.46) and idealized case $\alpha = 1/2$.

The level spectrum (16.79) simply results as:

$$
\begin{aligned}
E(\alpha = 1/2) &= \hbar\omega_0 \left( n_1 + n_2 + n_3 + \frac{3}{2} - \frac{1}{2}\delta_{n_1 0} \right) \\
&\quad + B\hbar\omega_0 \left( \frac{n_1!}{\Gamma(1/2 + n_1)} - \frac{1}{\Gamma(1/2)}\delta_{n_1 0} \right) \qquad (16.80) \\
&= \hbar\omega_0 \left( n_1 + n_2 + n_3 + \frac{3}{2} - \frac{1}{2}\delta_{n_1 0} \right) \\
&\quad + \frac{B\hbar\omega_0}{\sqrt{\pi}} \left( \frac{(2n_1)!!}{(2n_1 - 1)!!} - \delta_{n_1 0} \right) \qquad (16.81)
\end{aligned}
$$

where !! denotes the double factorial.

On the left side of Fig. 16.5 this spectrum is plotted in units $[\hbar\omega_0]$. Single levels are labelled according to the Nilsson-scheme and multiplicities are given in brackets. For small fractional field strength $B$ the resulting spectrum exactly follows the schematic level diagram of a phenomenological shell model with spin-orbit term, as demonstrated e.g. by Goeppert–Mayer [Goeppert-Mayer(1949)].

A small deviation from the ideal $\alpha = 1/2$ value reproduces the experimental spectra accurately:

For $\alpha = 0.48$ the resulting level spectrum is given on the right hand side of Fig. 16.5. Obviously there is an interference of two effects:

First, for $\alpha \neq 1/2$ now the degenerated levels of both magic sets split up and second the fractional magnetic field $B$ acts on the subset $n_{\text{magic }1}$.

For $B \approx 0.25$ the spectrum may be directly compared with the spherical Nilsson level scheme. For example, for neutrons between $82 \leq N \leq 126$ this level scheme is given as $2f\frac{7}{2}$, $1h\frac{9}{2}$, $1i\frac{13}{2}$, $3p\frac{3}{2}$, $2f\frac{5}{2}$, $3p\frac{1}{2}$, see e.g. results of [Scharnweber(1970)], which corresponds to a sequence of sub-shells at $90, 100, 114, 118, 124, 126$. This sequence is correctly reproduced with the extended Caputo–Riemann–Riemann symmetric rotor.

With the parameter set, which is obtained by a fit with the experimental

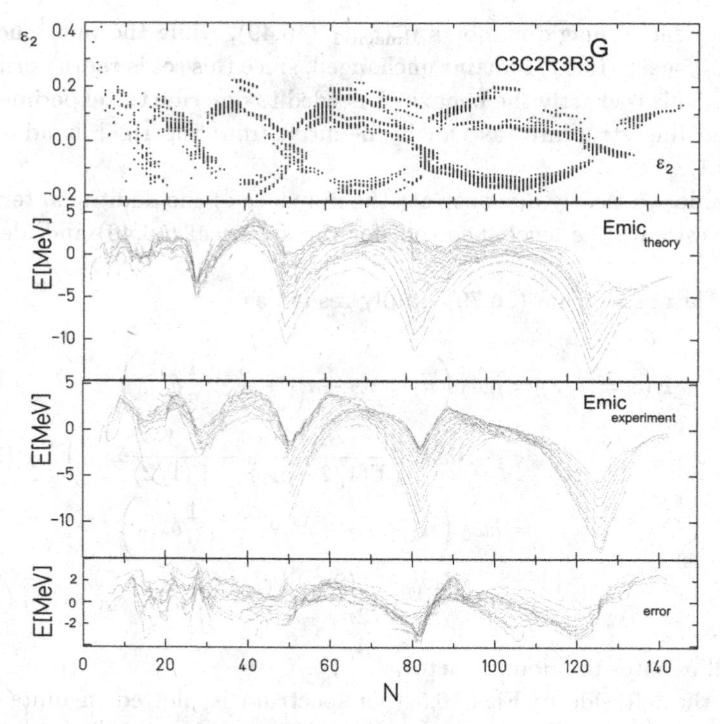


Fig. 16.6. Comparison of calculated $E_{\mathrm{mic}}$ from the extended Caputo–Riemann–Riemann symmetric rotor (16.79) with the parameter set (16.82)–(16.85), minimized with respect to $\epsilon_2$ with the experimental masses from Audi [Audi(2003)] as a function of $N$. From top to bottom the minimized $\epsilon_2$ values, theoretical and experimental masses and error in $[MeV]$ are plotted.

masses of Ca-, Sn- and Pb-isotopes

$$\hbar\omega_0 = 28A^{-\frac{1}{3}}[MeV] \tag{16.82}$$

$$\alpha_Z = \begin{cases} 0.480 + 0.00022\,Z & Z > 50 \\ 0.324 + 0.00332\,Z & Z \leq 50 \end{cases} \tag{16.83}$$

$$\alpha_N = \begin{cases} 0.446 + 0.00022\,N & N > 29 \\ 0.356 + 0.00332\,N & N \leq 29 \end{cases} \tag{16.84}$$

$$B = \begin{cases} 0 & Z < 40 \\ 0.0235\,Z - 0.94 & 40 \leq Z < 50 \\ 0.235 & Z \geq 50 \end{cases} \tag{16.85}$$

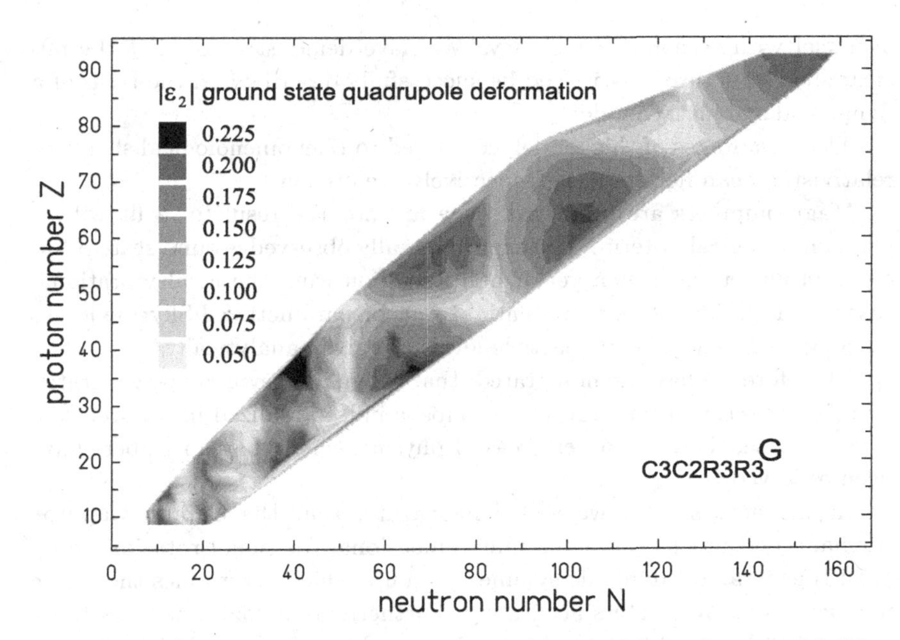


Fig. 16.7. Absolute value of the ground state quadrupole deformations $\epsilon_2$ minimizing the total energy for the extended Caputo–Riemann–Riemann symmetric rotor (16.79) with parameter set (16.82)–(16.85) for even-even nuclei. Gray levels indicate the increasing deformation from white ($|\epsilon_2| \leq 0.05$) to black ($|\epsilon_2| \geq 0.225$).

the experimental masses are reproduced with an rms-error of 1.7[MeV]. Results are given in Figs. 16.6 and 16.7.

The deformation parameters, obtained by minimization of the total energy, are to a large extent consistent with values given in [Möller(1995)] e.g. for $^{264}\mathrm{Hs}_{108}$ we obtain $\epsilon_2 = 0.22$, which conforms with Möller's ($\epsilon_2 = 0.2$) and Rutz's results [Rutz(1997)]. However, there occur discrepancies mostly for exotic nuclei. For example our calculations determine the nucleus $^{42}\mathrm{Si}$ to be almost spherical ($\epsilon_2 = -0.03$), which conforms with recent experimental findings [Fridmann(2005)], while Möller predicts a definitely oblate shape ($\epsilon_2 = -0.3$).

Finally, defining a nucleus with $\epsilon_2 > 0.05$ as prolate and with $\epsilon_2 < -0.05$ as oblate the amount of prolate shapes is about 74% of all deformed nuclei. This is close to the value of 82% [Tajima(2001)], obtained with the Nilsson model using the standard parameters.

Summarizing the results presented, the proposed extended Caputo–Riemann–Riemann symmetric rotor describes the ground state properties of nuclei with reasonable accuracy. We have demonstrated, that the nuclear shell structure may indeed be successfully described on the basis of a dynamical symmetry model.

The advantages of this model, compared to phenomenological shell and relativistic mean field models respectively are obvious:

Magic numbers are predicted, they are not the result of a fit with a phenomenological $ls$-term. The experimentally observed ground state properties of nuclei occur at a very small deviation from the ideal vibrational case $\alpha = 1/2$. There are no potential-terms or parametrized Skyrme-forces involved and finally, single particle levels are given analytically.

Therefore we have demonstrated, that a dynamic symmetry, generated by mixed fractional type rotation groups is indeed realized in nature.

Of course, there are other areas of physics, where magic numbers have been observed.

In the next section we will demonstrate, that the Caputo–Caputo–Riemann decomposition of the nine dimensional fractional rotation group (16.13) generates a dynamic symmetry group, which determines the magic numbers in metal clusters accurately. Furthermore a comparison with experimental data will lead to the conclusion, that a fractional phase transition occurs near cluster size $200 \leq \mathcal{N} \leq 300$.

## 16.7 Magic numbers of electronic clusters: the nine dimensional fractional Caputo–Caputo–Riemann symmetric rotor

Since 1984, an increasing amount of experimental data [Knight(1984), Martin(1991)] confirms an at first unexpected shell structure in fermion systems, realized as magic numbers in metal clusters.

The observation of varying binding energy of the valence electron, moving freely in a metallic cluster, has initiated the development of several theoretical models. Besides ab initio calculation, the most prominent representatives are the jellium model [Brack(1993)] and, in analogy to methods already in use in nuclear physics, phenomenological shell models with modified potential terms like the Clemenger–Nilsson model or deformed Woods–Saxon potential [de Heer(1993), Moriarty(2001)].

Although these models describe the experimental data with reasonable accuracy, they do not give a theoretical explanation for the observed sequence of magic numbers. Therefore the problem of a theoretical foundation of the magic numbers is still an open question.

In the preceding sections we have demonstrated, that the magic numbers in atomic nuclei are the result of a fractional dynamic symmetry, which is determined by a specific decomposition of the nine dimensional mixed fractional rotation group.

On the basis of this encouraging result, we will demonstrate, that a more fundamental understanding of magic numbers found for metal clusters may be achieved applying an alternative decomposition of the same nine dimensional fractional rotation group.

According to the group chain (16.13) we associate the Hamiltonian:

$$H = \frac{\omega_1}{\hbar}\,{}_{\mathrm{C}}L_1^2(\alpha) + \frac{\omega_2}{\hbar}\,{}_{\mathrm{C}}L_2^2(\alpha) + \frac{\omega_3}{\hbar}\,{}_{\mathrm{R}}L_3^2(\alpha) \tag{16.86}$$

with the free parameters $\omega_1, \omega_2, \omega_3$ and the basis is $|L_1 M_1 L_2 M_2 L_3 M_3\rangle$.

Imposing the same symmetries as in Sec. 16.3 we obtain the corresponding level spectrum:

$$\begin{aligned} E(\alpha) &= \hbar\omega_1\,{}_{\mathrm{C}}[2n_1]_{\mathrm{C}}[2n_1+1] + \hbar\omega_2\,{}_{\mathrm{C}}[2n_2]_{\mathrm{C}}[2n_2+1] \\ &\quad + \hbar\omega_3\,{}_{\mathrm{R}}[2n_3]_{\mathrm{R}}[2n_3+1] \qquad (16.87) \\ &= \sum_{i=1}^{3} \hbar\omega_i \frac{\Gamma(1+(2n_i+1)\alpha)}{\Gamma(1+(2n_i-1)\alpha)} \\ &\quad - \delta_{n_1 0}\hbar\omega_1 \frac{\Gamma(1+\alpha)}{\Gamma(1-\alpha)} - \delta_{n_2 0}\hbar\omega_2 \frac{\Gamma(1+\alpha)}{\Gamma(1-\alpha)} \qquad (16.88) \\ &\qquad n_1, n_2, n_3 = 0, 1, 2, \end{aligned}$$

on a basis $|2n_1 2n_1 2n_2 2n_2 2n_3 2n_3\rangle$.

This is the major result of our derivation. We call this model the Caputo–Caputo–Riemann symmetric rotor. What makes this model remarkable once again is its behaviour in the vibrational region near the semiderivative $\alpha = 1/2$.

In Fig. 16.8 we have plotted the energy levels in the vicinity of $\alpha \approx 1/2$ for the spherical case (16.46).

For the idealized spherical case $\alpha = 1/2$, using the relation $\Gamma(1+z) = z\Gamma(z)$ the level spectrum (16.87) is simply given by:

$$E(\alpha = 1/2) = \hbar\omega_0 \left( n_1 + n_2 + n_3 + \frac{3}{2} - \frac{1}{2}\delta_{n_1 0} - \frac{1}{2}\delta_{n_2 0} \right) \tag{16.89}$$

In order to determine the multiplets of (16.89), we first recall the properties of the well known spectrum of the spherical 3-dimensional harmonic oscillator.

$$E_{\mathrm{HO}} = \hbar\omega_0(n_1 + n_2 + n_3 + 3/2) \tag{16.90}$$

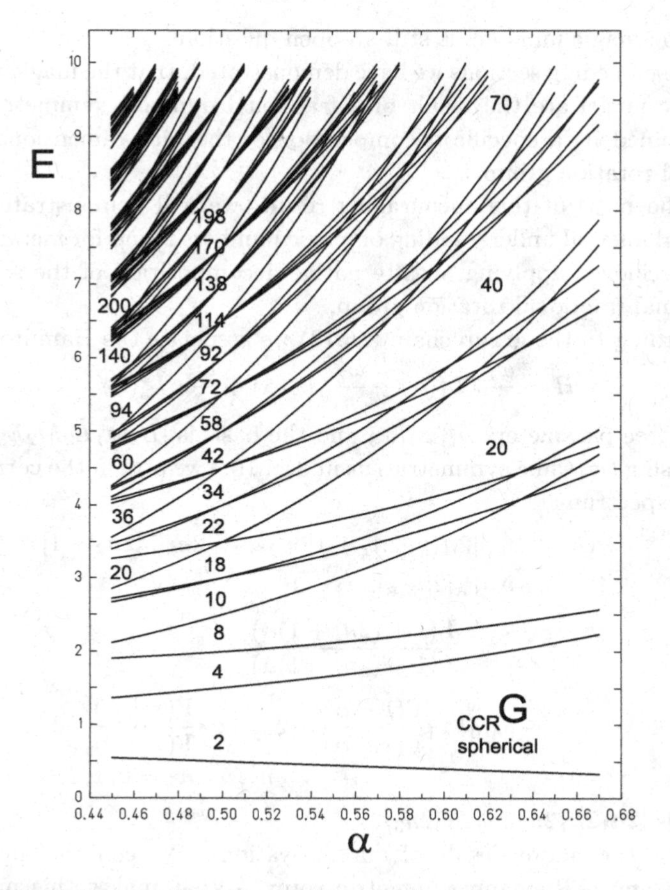


Fig. 16.8. Level diagram of the Caputo–Caputo–Riemann symmetric rotor (16.87) for the spherical case in the vicinity of $\alpha \approx 1/2$.

We introduce the quantum number $N$ as

$$N = n_1 + n_2 + n_3 \tag{16.91}$$

Assuming a two-fold spin degeneracy of the energy levels, we obtain the multiplets of the spherical three dimensional harmonic oscillator with magic numbers for

$$n_{\text{HO}}(N) = \frac{1}{3}(N+1)(N+2)(N+3), \qquad N = 0, 1, 2, 3, ... \tag{16.92}$$

$$= 2, 8, 20, 40, 70, 112, 168, 240, ... \tag{16.93}$$

at energies

$$E(N)_{\text{HO}} = \hbar\omega_0(N + 3/2) \tag{16.94}$$

For the multiplets of (16.89) we distinguish two different sets of magic numbers, which we label with $N_1$ and $N_2$:

For $n_1 = 0$ and $n_2 = 0$ the multiplicity of a harmonic oscillator shell for $N$ at energy (16.94) is increased by exactly one state, the $|0000\,N{+}1\,N{+}1\rangle$ state, which originates from the $N+1$ shell. Therefore we obtain a first set ${}_{\text{CCR}}n_{\text{magic 1}}$ of magic numbers (including the state $|000000\rangle$ for which we assign $N = -1$):

$$
\begin{aligned}
{}_{\text{CCR}}n_{\text{magic 1}}(N_1) &= n_{\text{HO}}(N_1 - 1) + 2, \qquad N_1 = 0, 1, 2, 3, \ldots && (16.95)\\
&= \frac{1}{3} N_1(N_1 + 1)(N_1 + 2) + 2 && (16.96)\\
& \quad 2, 4, 10, 22, 42, 72, 114, 170, 242, 332, 442, \ldots && (16.97)
\end{aligned}
$$

at energies

$$
E(N_1)_{\text{CCR}}n_{\text{magic 1}} = \hbar\omega_0(N_1 + 1/2) \tag{16.98}
$$

In addition, for $n_1 = 0$, which corresponds to the $|00\,2n_2 2n_2\,2n_3 2n_3\rangle$ states and $n_2 = 0$ respectively, which corresponds to the $|2n_1 2n_1\,00\,2n_3 2n_3\rangle$ states with a $\sum_{n=1}^{N+1} 2 = 2(N+1)$-fold multiplicity for each set $n_1 = 0$ and $n_2 = 0$ we obtain a second set ${}_{\text{CCR}}n_{\text{magic 2}}$ of magic numbers

$$
\begin{aligned}
{}_{\text{CCR}}n_{\text{magic 2}}(N_2) &= {}_{\text{CCR}}n_{\text{magic 1}}(N_2 + 1) + 4(N_2 + 1) && (16.99)\\
& \qquad N_2 = 0, 1, 2, 3, \ldots \\
&= \frac{1}{3}(N_2 + 1)(N_2 + 2)(N_2 + 3) + 2 + 4(N_2 + 1) && (16.100)\\
&= 8, 18, 34, 58, 92, 138, 198, 274, \ldots && (16.101)
\end{aligned}
$$

at energies

$$
E(N_2)_{\text{CCR}}n_{\text{magic 2}} = \hbar\omega_0(N_2 + 2) \tag{16.102}
$$

In Fig. 16.8, the single particle levels are plotted. A remarkable feature is the dominant influence of the $|0000NN\rangle$ state. For $\alpha < 0.5$ the harmonic oscillator type magic numbers die out. As a consequence, for $\alpha \approx 0.48$ the set of magic numbers ${}_{\text{CCR}}n_{\text{magic 2}}$ is shifted by 1, which leads to the series $2, 4, 9, 19, 35, 59, 93, 139, \ldots$. For $\alpha \approx 0.46$ the $|0000NN\rangle$ state has completely reached the ${}_{\text{CCR}}n_{\text{magic 2}}$ multiplet, which in a stable series of magic numbers at $2, 4, 10, 20, 36, 60, 94, 140, 200, \ldots$.

On the other hand for $\alpha > 0.55$ the levels are rearranged to form the set of magic numbers of the harmonic oscillator.

We conclude, that the Caputo–Caputo–Riemann symmetric rotor predicts a well defined set of magic numbers. This set is a direct consequence of the underlying dynamic symmetries of the three fractional rotation groups involved. It is indeed remarkable, that the same set of magic numbers is realized in nature as electronic magic numbers in metal clusters.

In the next section we will demonstrate, that the proposed analytical model is an appropriate tool to describe the shell correction contribution to the total binding energy of metal clusters.

## 16.8 Binding energy of electronic clusters

We will use the Caputo–Caputo–Riemann symmetric rotor (16.87) as a dynamic shell model for a description of the microscopic part of the total energy binding energy $E_{\text{tot}}$ of the metal cluster:

$$E_{\text{tot}} = E_{\text{macroscopic}} + E_{\text{microscopic}} \tag{16.103}$$

$$= E_{\text{macroscopic}} + \delta U \tag{16.104}$$

where $\delta U$ denotes the shell-correction contributions.

To make our argumentation as clear as possible, we will restrict our investigation to the spherical configuration, which will allow to discuss the main features of the proposed model in a simple context. We will compare our results with calculations for the most prominent metal cluster, the sodium (Na) cluster. From experimental data [Knight(1984)], [Bjornholm(1990)], [Martin(1991)], the following sequence of magic numbers is deduced:

$$\begin{aligned} n_{\text{magic Na}} = \{&2, 8, 20, 40, 58, 92, 138, 198 \pm 2, 263 \pm 5, 341 \pm 5, \\ &443 \pm 5, 557 \pm 5, 700 \pm 15, 840 \pm 15, 1040 \pm 20, \\ &1220 \pm 20, 1430 \pm 20\} \end{aligned} \tag{16.105}$$

For a graphical representation of the experimental magic numbers we introduce the two quantities:

$$\Theta = i/(n_{\text{magic Na}}(i+1) - n_{\text{magic Na}}(i)) \tag{16.106}$$

$$\omega = n_{\text{magic Na}}(i+1) - n_{\text{magic Na}}(i) \tag{16.107}$$

where $i$ denotes the array-index in (16.105).

Interpreting $\Theta$ as a moment of inertia and $\omega$ a rotational frequency, Fig. 16.9 is a back-bending plot of the experimental magic numbers.

We distinguish three different regions of magic numbers:

For cluster size $\mathcal{N} < 200$ the plot shows a typical rotor spectrum. In the region $200 < \mathcal{N} < 300$ a typical back-bending phenomenon is observed.

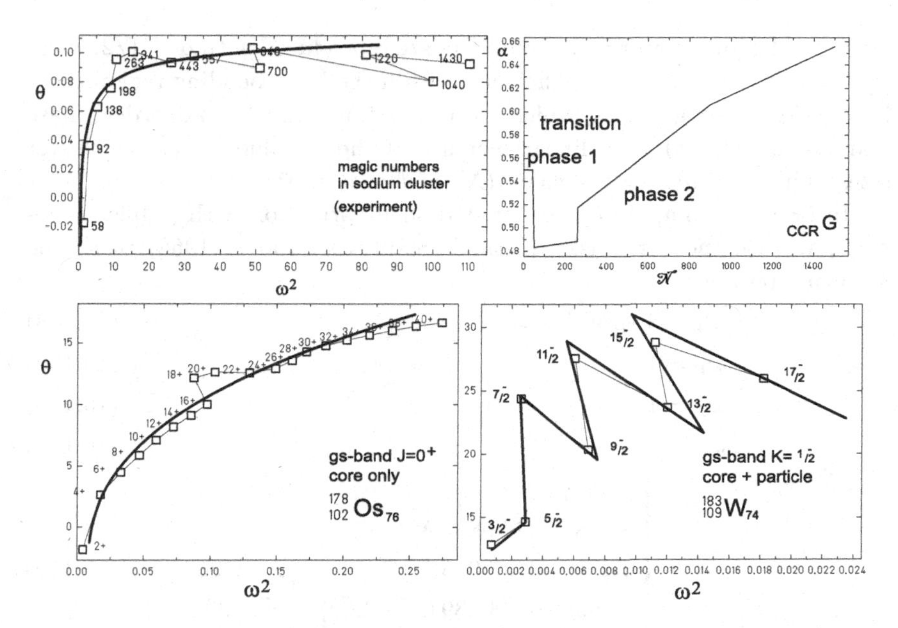


Fig. 16.9. Back-bending plots of experimentally determined magic numbers for $(\mathrm{Na})_{\mathcal{N}}$ clusters from [Knight(1984), Bjornholm(1990), Martin(1991)], ground state excitation spectrum for ${}^{178}_{102}\mathrm{Os}_{76}$ from Chapter 13.7 and ground state excitation spectrum of ${}^{183}_{109}\mathrm{W}_{74}$ from Chapter 14.4. Squares indicate the experimental values. The sequence of experimentally observed magic numbers may be categorized for $\mathcal{N} < 200$ to be equivalent to an excitation spectrum of purely rotational type, the thick line indicating a fit with a fractional rigid rotor spectrum, for $200 \leq \mathcal{N} \leq 300$ a region of back-bending type (compare with the plotted ground state band spectrum for ${}^{178}_{102}\mathrm{Os}_{76}$) and finally, for $\mathcal{N} > 300$ a region with almost constant $\theta$ and a behaviour similar to excitation spectra of ug-nuclei (compare with the plotted excitation spectrum of ${}^{183}_{109}\mathrm{W}_{74}$). In the upper right the corresponding proposed $\alpha(N)$ from (16.112) is plotted.

For illustrative purposes in Fig. 16.9 the same phenomenon is documented within the framework of nuclear physics for the ground state rotation spectrum of ${}^{178}_{102}\mathrm{Os}_{76}$. For $\mathcal{N} > 300$ the moment of inertia becomes nearly constant and the graph may be compared with the rotational $K = \frac{1}{2}$ band of the ug-nucleus ${}^{183}_{109}\mathrm{W}_{74}$, which is a typical example of a core plus single particle motion in nuclear physics.

These different structures in the sequence of electric magic numbers are reflected in the choice of the fractional derivative coefficient $\alpha$. For $\mathcal{N} < 200$, $\alpha$ shows a simple behaviour similar to the case of magic nucleon numbers, it varies in the vicinity of $\alpha \approx 1/2$. For the special case of sodium clusters, the lowest four magic numbers are reproduced with $\alpha > 1/2$, while up to $\mathcal{N} = 198$ $\alpha < 1/2$ is sufficient. Within the back-bending region there is a sudden change in $\alpha$, which we call a fractional second order phase transition, followed by a linear increase of the $\alpha$ value for larger cluster sizes. The resulting dependence $\alpha(\mathcal{N})$ is shown in Fig. 16.9.

In order to compare our calculated shell correction with published results, we use the Strutinsky method [Strutinsky(1967b, 1968)] with the following parameters:

$$\hbar\omega_0 = 3.96\mathcal{N}^{-\frac{1}{3}}\,[\text{eV}] \tag{16.108}$$

$$\omega_1 = 1 \tag{16.109}$$

$$\omega_2 = 1 \tag{16.110}$$

$$\omega_3 = 1 \tag{16.111}$$

$$\alpha = \begin{cases} 0.55 & \mathcal{N} < 43 \\ 0.908 - 0.000834\,\mathcal{N} & \mathcal{N} < 51 \\ 0.482 + 0.000025\,\mathcal{N} & \mathcal{N} < 260 \\ 0.069 + 0.000139\,(\mathcal{N} - 260) & \mathcal{N} < 900 \\ 0.062 + 0.000083\,(\mathcal{N} - 260) & \mathcal{N} \geq 900 \end{cases} \tag{16.112}$$

$$\gamma = 1.1\,\hbar\omega_0\big({}_{\mathrm{R}}[\mathcal{N}^{1/3} + 1] - {}_{\mathrm{R}}[\mathcal{N}^{1/3}]\big)^3 \tag{16.113}$$

$$m = 4 \tag{16.114}$$

(16.113) follows from the plateau condition $\partial U/\partial\gamma = 0$ and (16.114) is the order of included Hermite polynomials for the Strutinsky shell correction method.

In Fig. 16.10 the resulting shell correction $\delta U$ is plotted. Magic numbers are reproduced correctly within the experimental errors. Furthermore we obtain a nearly quantitative agreement with published results for the shell correction term obtained e.g. with the spherical Woods–Saxon potential [Bjornholm(1990), Nishioka(1990)].

Summarizing the results presented so far, the proposed Caputo–Caputo–Riemann symmetric rotor describes the magic numbers and microscopic part of the total binding energy for metal clusters with reasonable accuracy. We have demonstrated, that the cluster shell structure may indeed be successfully described on the basis of a dynamical symmetry model.

The behaviour of metallic clusters is dominated by electromagnetic forces, while in nuclei the long range part of strong forces is important. Therefore it has been shown that two out of the four different decompositions (16.11)–(16.14) of the nine dimensional fractional rotation group

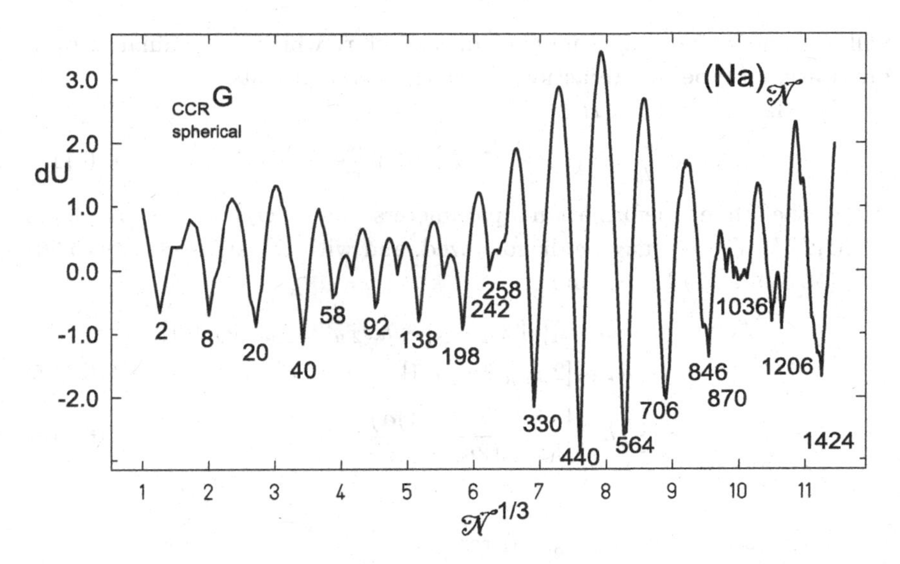


Fig. 16.10. Calculated shell correction $dU$ as a function of cluster size $\mathcal{N}$ for the Caputo–Caputo–Riemann fractional symmetric spherical rotor with parameter set (16.108)–(16.114). This graph should be directly compared with [Bjornholm(1990)].

generate dynamical symmetries, which dominate the ground state properties of atomic nuclei and metallic clusters.

## 16.9 Metaphysics: magic numbers for clusters bound by weak and gravitational forces respectively

We have demonstrated, that the Caputo–Caputo–Riemann rotor (16.87) correctly determines the magic numbers of metal clusters and that the Caputo–Riemann–Riemann rotor (16.43) is an appropriate tool to describe the ground state properties of nuclei with reasonable accuracy. Both models differ only in the mixing ratio of fractional derivatives. The phenomena described, differ only in the interaction type of the constituents which build up the cluster. The behaviour of metallic clusters is dominated by electromagnetic forces, while in nuclei the long range part of strong forces is important.

Therefore we postulate, that the group decomposition (16.11)

$$ {}_{\mathrm{RRR}}G \supset {}_{\mathrm{R}}SO^{\alpha}(3) \supset {}_{\mathrm{R}}SO^{\alpha}(3) \supset {}_{\mathrm{R}}SO^{\alpha}(3) \qquad (16.115) $$

will determine the magic number of a cluster, which is dominated by a gravitational type of interaction between its constituents.

The Hamiltonian ${}_{\mathrm{RRR}}H$

$$ {}_{\mathrm{RRR}}H = \frac{\omega_1}{\hbar}\,{}_{\mathrm{R}}L_1^2(\alpha) + \frac{\omega_2}{\hbar}\,{}_{\mathrm{R}}L_2^2(\alpha) + \frac{\omega_3}{\hbar}\,{}_{\mathrm{R}}L_3^2(\alpha) \tag{16.116} $$

with the free deformation parameters $\omega_1$, $\omega_2$,$\omega_3$ on a basis $|L_1M_1L_2M_2L_3M_3\rangle$ may be diagonalized, and with the symmetries (16.38) and (16.39), a level spectrum

$$ \begin{aligned} E(\alpha) &= \hbar\omega_1\,{}_{\mathrm{R}}[2n_1]_{\mathrm{R}}[2n_1+1] + \hbar\omega_2\,{}_{\mathrm{R}}[2n_2]_{\mathrm{R}}[2n_2+1] \\ &\quad + \hbar\omega_3\,{}_{\mathrm{R}}[2n_3]_{\mathrm{R}}[2n_3+1] && (16.117) \\ &= \sum_{i=1}^{3} \hbar\omega_i \frac{\Gamma(1+(2n_i+1)\alpha)}{\Gamma(1+(2n_i-1)\alpha)} && (16.118) \\ &\qquad\qquad\qquad n_1, n_2, n_3 = 0, 1, 2, \end{aligned} $$

on a basis $|2n_1 2n_1 2n_2 2n_2 2n_3 2n_3\rangle$ results.

For the idealized spherical case $\alpha = 1/2$ this spectrum is simply given by:

$$ E(\alpha = 1/2) = \hbar\omega_0\left(n_1 + n_2 + n_3 + \frac{3}{2}\right) \tag{16.119} $$

which is the spectrum of the deformed harmonic oscillator. In the spherical case, magic numbers are determined by:

$$ \begin{aligned} n_{\mathrm{RRR}} &= \frac{1}{3}(N+1)(N+2)(N+3), \qquad N = 0,1,2,3,\ldots && (16.120) \\ &= 2, 8, 20, 40, 70, 112, 168, 240, \ldots && (16.121) \end{aligned} $$

at energies

$$ E(N)_{\mathrm{RRR}} = \hbar\omega_0(N + 3/2) \tag{16.122} $$

This result may be compared with solutions for an independent particle shell model, where the potential is determined by a uniformly distributed gravitational charge (mass) distribution $\rho(r) = q/V$ inside a sphere. This potential is given by

$$ \begin{aligned} V(r) &= \int\!\!\int\!\!\int \frac{\rho(r')}{|r - r'|} d^3r' && (16.123) \\ &= q\left(\frac{r^2}{2R_0^3} - \frac{3}{2R_0}\right), \qquad r < R_0 && (16.124) \end{aligned} $$

and leads to a radial Schrödinger equation for the harmonic oscillator.

Therefore we are led to the prediction, that for microscopic clusters with gravitational type of interaction of the constituents there will be variations in the binding energy per mass unit according to (16.120). We cannot predict the value of the mass unit, but since the magnitude of shell corrections for metal clusters is of order eV and for nuclear shell corrections of order MeV, we assume the magnitude of shell corrections for gravity dominated clusters to be of order TeV, which amounts about 1000 proton masses or $10^{-23}$[kg].

Consequently we are left with the fourth decomposition of the nine dimensional fractional rotation group (16.14)

$$_{\text{CCC}}G \supset {}_{\text{C}}SO^{\alpha}(3) \supset {}_{\text{C}}SO^{\alpha}(3) \supset {}_{\text{C}}SO^{\alpha}(3) \tag{16.125}$$

The Hamiltonian $_{\text{CCC}}H$

$$_{\text{CCC}}H = \frac{\omega_1}{\hbar}\,{}_{\text{C}}L_1^2(\alpha) + \frac{\omega_2}{\hbar}\,{}_{\text{C}}L_2^2(\alpha) + \frac{\omega_3}{\hbar}\,{}_{\text{C}}L_3^2(\alpha) \tag{16.126}$$

with the free deformation parameters $\omega_1$, $\omega_2$, $\omega_3$ on a basis $|L_1 M_1 L_2 M_2 L_3 M_3\rangle$ may be diagonalized, and with the symmetries (16.38) and (16.39), a level spectrum

$$E(\alpha) = \hbar\omega_1\,{}_{\text{C}}[2n_1]_{\text{C}}[2n_1+1] + \hbar\omega_2\,{}_{\text{C}}[2n_2]_{\text{C}}[2n_2+1] + \hbar\omega_3\,{}_{\text{C}}[2n_3]_{\text{C}}[2n_3+1] \tag{16.127}$$

$$= \sum_{i=1}^{3} \hbar\omega_i \frac{\Gamma(1+(2n_i+1)\alpha)}{\Gamma(1+(2n_i-1)\alpha)} - \delta_{n_1 0}\hbar\omega_1 \frac{\Gamma(1+\alpha)}{\Gamma(1-\alpha)} - \delta_{n_2 0}\hbar\omega_2 \frac{\Gamma(1+\alpha)}{\Gamma(1-\alpha)} - \delta_{n_3 0}\hbar\omega_3 \frac{\Gamma(1+\alpha)}{\Gamma(1-\alpha)} \tag{16.128}$$

$$n_1, n_2, n_3 = 0, 1, 2, \ldots$$

on a basis $|2n_1 2n_1 2n_2 2n_2 2n_3 2n_3\rangle$ results.

We call this model the Caputo–Caputo–Caputo symmetric rotor. For the idealized spherical case $\alpha = 1/2$ this spectrum is simply given by:

$$E(\alpha = 1/2) = \hbar\omega_0 \left( n_1 + n_2 + n_3 + \frac{3}{2} - \frac{1}{2}\delta_{n_1 0} - \frac{1}{2}\delta_{n_2 0} - \frac{1}{2}\delta_{n_3 0} \right) \tag{16.129}$$

We obtain a first set $_{\text{CCC}}n_{\text{magic 1}}$ of magic numbers

$$_{\text{CCC}}n_{\text{magic 1}} = n_{\text{HO}} + 6, \qquad N = 0, 1, 2, 3, \ldots \tag{16.130}$$

$$= \frac{1}{3}(N+1)(N+2)(N+3) + 6 \tag{16.131}$$

$$8, 14, 26, 46, 76, 118, 174, 246, 336, 446, \ldots \tag{16.132}$$

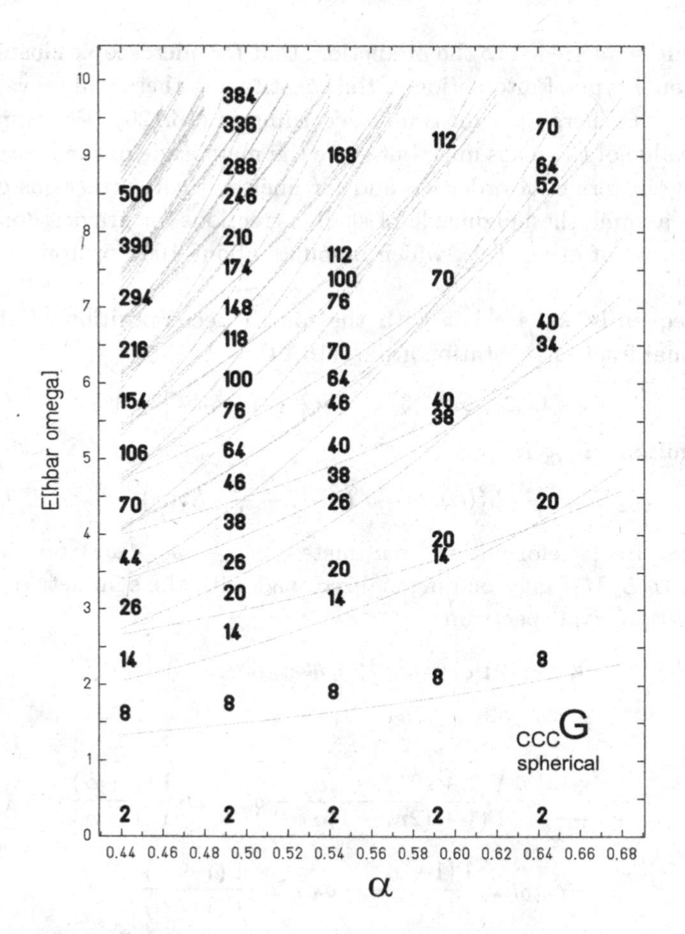


Fig. 16.11. Level diagram of the Caputo–Caputo–Caputo symmetric rotor for the spherical case in the vicinity of $\alpha \approx 1/2$.

at energies

$$E(N)_{\text{CCC}}n_{\text{magic 1}} = \hbar\omega_0(N + 3/2) \tag{16.133}$$

In addition we obtain a second set ${}_{\text{CCC}}n_{\text{magic 2}}$ of magic numbers

$$\begin{aligned} {}_{\text{CCC}}n_{\text{magic 2}} &= {}_{\text{CCC}}n_{\text{magic 1}} + 6N, \qquad N = 1, 2, 3, ... && (16.134)\\ &= \frac{1}{3}(N+1)(N+2)(N+3) + 6(N+1) && (16.135)\\ &= 20, 38, 64, 100, 148, 210, 288, ... && (16.136) \end{aligned}$$

at energies

$$E(N)_{\mathrm{CCC}} n_{\mathrm{magic}\ 2} = \hbar\omega_0 (N+1) \tag{16.137}$$

Finally, the state $|000000\rangle$ with a two-fold multiplicity has energy $E = 0$ and therefore does fit into one of the two derived series.

Consequently we are led to the prediction, that for microscopic clusters with weak interaction type of the constituents there will be variations in the binding energy per charge unit according to (16.134). We estimate the order of magnitude of shell corrections for weak interaction dominated clusters to be $10^{-6}$[eV], which seems reasonable for e.g. a neutrino-cluster.

Summarizing the results presented in this section, we have associated the four different decompositions (16.11)–(16.14) of the nine dimensional mixed fractional rotation group with the four fundamental types of interaction found in nature. We found common aspects determining the magic numbers for each group. There are always two different sets of magic numbers, one set is a shifted harmonic oscillator set, the other set is specific to the group considered. For the spherical and idealized case $\alpha = 1/2$ the four different sequences of magic number sets are simply the result of the presence or absence of a Kronecker-delta.

Our investigations lead to the conclusion, that mixed fractional derivative type field theories may play an important role in a unified theory including all four fundamental interactions.

# Chapter 17

# Fractors: Fractional Tensor Calculus

*No specific theory of fractional tensor calculus exists up to now. In the following we propose notations, which may be of use for a formulation of such a theory. Especially we propose to interpret the fractional derivative parameter $\alpha$ as an extension of tensor calculus on a level similar to spin, which leads to the definition of spinors [Schmutzer(1968)].*

*Consequently the objects considered in fractional tensor calculus may be called fractors.*

## 17.1 Covariance for fractional tensors

Especially the fractional derivative and in general the quantities being investigated, are elements of a direct product space of the discrete coordinate space, whose discrete coordinate indices are denoted by $\mu, \nu, \ldots = 0, \ldots, 3$ and the continuous space of fractional derivatives, which are labelled with $\alpha, \beta, \ldots \in \mathbb{R}$. These are real numbers, which in the case $\alpha < 0$ or $\beta < 0$ are interpreted as a fractional integral of order $\alpha$ and $\beta$ respectively.

For the covariant fractional derivatives of order $\alpha$ we introduce the short hand notations

$$\frac{\partial^\alpha}{\partial (x^\mu)^\alpha} f = \partial^\alpha_\mu f = f^\alpha_{|\mu} \tag{17.1}$$

Einstein's summation convention for tensors e.g.

$$\partial_\mu = g_{\mu\nu}\partial^\nu = \sum_{\nu=0}^{3} g_{\mu\nu}\partial^\nu \tag{17.2}$$

is extended to the continuous case for the fractional derivative coefficients

$$\partial_\alpha = g_{\alpha\beta}\partial^\beta = \int_{\beta=-\infty}^{+\infty} d\beta \hat{g}(\alpha, \beta)\partial^\beta \tag{17.3}$$

where $\hat{g}(\alpha, \beta)$ is an appropriately chosen function of the two variables $\alpha, \beta$.

Hence we define the generalized metric tensor

$$\eta^{\mu\nu}_{\alpha\beta} = \text{diag}\{-1, 1, 1, 1\}\delta(\alpha - \beta) \tag{17.4}$$

where $\delta(\alpha - \beta)$ is the Dirac-delta function for raising and lowering of indices.

Therefore the fractional derivative objects are formally described in a manner similar to the case of spinor objects. A quantity like e.g. $F^{\alpha}_{\mu\nu}$ transforms as a tensor of rank 2 in coordinate space and as a vector in the sense of (17.3) in fractional derivative space, which in analogy to spinors we may call a fractor of rank one.

With these definitions, the contravariant fractional derivative follows as:

$$\partial^{\mu}_{\alpha} f = f^{|\mu}_{\alpha} = \eta^{\mu\nu}_{\alpha\beta}\partial^{\beta}_{\nu} f \tag{17.5}$$

A scalar quantity like the Lagrangian density of a fractional electromagnetic field is then generated by contracting the indices in the usual way:

$$\mathcal{L}_{\text{fEM}} = -\frac{1}{4} F^{\alpha}_{\mu\nu} F^{\mu\nu}_{\alpha} \tag{17.6}$$

## 17.2 Singular fractional tensors

Finally, we have to resolve an ambiguity, which is specific to fractional tensor calculus. For $\alpha = 0$ the use of the fractional derivative introduces an index and therefore

$$\partial^{0}_{\mu} f = f \tag{17.7}$$

obviously violates the principle of covariance, since the tensor rank differs for left and right hand side of (17.7). Consequently we introduce the Kronecker-delta $\delta^{\mu}_{0}$:

$$\delta^{\mu}_{0}\partial^{0}_{\mu} f = \delta^{0}_{\mu}\partial^{\mu}_{0} f = f \tag{17.8}$$

and perform a pseudo-summation, which is reduced to a single term. We introduce the abbreviations

$$\delta^{\mu}_{0}\partial^{0}_{\mu} = \delta^{(\mu)} \tag{17.9}$$

$$\delta^{0}_{\mu}\partial^{\mu}_{0} = \delta_{(\mu)} \tag{17.10}$$

and call $(\mu)$ a pseudo-index, where summation is performed, but the tensor properties in fractional coordinate space are not affected.

The Kronecker-delta may be generalised to

$$\delta_{(\mu)}{}^{\tau} \partial^{\alpha}_{\tau} f = \partial^{\alpha}_{(\mu)} f \tag{17.11}$$

where $\partial^{\alpha}_{(\mu)} f$ transforms like a scalar in fractional coordinate space, but the derivative with respect to $\mu$ is performed. With this definition, the Kronecker-delta will be used like

$$\delta_{(\mu)}{}^{\tau\sigma} \partial^{\alpha}_{\tau} \partial^{-\beta}_{\sigma} f = \delta_{(\mu)}{}^{\tau} \partial^{\alpha-\beta}_{\tau} f = \partial^{\alpha-\beta}_{(\mu)} f \tag{17.12}$$

Hence introducing the Kronecker-delta extends manifest covariant tensor calculus in a reasonable way to fractional derivative tensors.

# Chapter 18

# Fractional Fields

*The dynamical behaviour of fields may be deduced with appropriately chosen equations of motion, which in general will be a complex system of interdependent components.*

*Introducing a Lagrangian density, which is a function of the fields and their derivatives allows for an elegant description from a generalized point of view, since it is a scalar quantity, from which the field equations may be deduced using a variation principle.*

*Hence a Lagrangian density serves as a starting point to model a dynamic field theory.*

*Classical examples for Lagrangian densities, which play an important role in physics include:*

- *The Lagrangian density of the electromagnetic field has two terms, one for the electric charge density $J_\mu$ coupled to an external potential $A_\mu$ and another term with the electromagnetic field strength tensor $F_{\mu\nu}$:*

$$\mathcal{L}_{EM} = J^\mu A_\mu - \frac{1}{4} F_{\mu\nu} F^{\mu\nu} \tag{18.1}$$

- *The Lagrangian density for a free Dirac-field is given by:*

$$\mathcal{L}_D = \bar{\psi}(i\gamma^\mu \partial_\mu + m)\psi \tag{18.2}$$

- *For the simplest version of quantum hadrodynamics of relativistic nuclear matter [Serot(1986)] the Walecka model is defined starting with a Lagrangian density for the nucleon ($\psi$), the neutral scalar ($\phi$) and the vector ($\omega^\mu$) mesonic fields:*

$$\begin{aligned} \mathcal{L}_W = & i\bar{\psi}\partial_\mu \gamma^\mu \psi - M\bar{\psi}\psi + \bar{\psi}(g_s\phi - g_v \omega_\mu \gamma^\mu)\psi \\ & + \frac{1}{2}[\partial_\mu \phi \partial^\mu \phi - m_s^2 \phi^2] - \frac{1}{4}\mathcal{F}^{\mu\nu}\mathcal{F}_{\mu\nu} + \frac{1}{2} m_v^2 \omega^2 \end{aligned} \tag{18.3}$$

*where $M$, $m_s$, $m_v$ are the masses of the nucleon, scalar and vector mesons, respectively; $g_s$, $g_v$ are the coupling constants and $\mathcal{F}^{\mu\nu} = \partial^\mu \omega^\nu - \partial^\nu \omega^\mu$.*

- *Last but not least, the Landau–Ginzburg model of superconductivity, which is one of the ancestors of the non-Abelian Higgs-model [Higgs(1964)] is based on a Lagrangian density of the form:*

$$\mathcal{L}_{LG} = \frac{1}{4} F^{\mu\nu} F_{\mu\nu} + (D_\mu \phi)(D^\mu \phi) + V(|\phi|) \tag{18.4}$$

*where the gauge covariant derivative $D_\mu$ is given as*

$$D_\mu \phi = \partial_\mu \phi - igA_\mu \phi \tag{18.5}$$

*and the potential $V(|\phi|)$ models a self-interaction e.g. $V(|\phi| = \phi^4)$.*

*In modern theory the Lagrangian density is the formal tool to model dynamic properties of fields.*

*Since we want to investigate the properties of fractional fields, a formulation in terms of fractional Lagrangian densities could be helpful and is a research area of increasing interest [Baleanu(2005), Agrawal(2007), El-Nabulsi(2008), Baleanu(2010)].*

*Fractional Lagrangian densities are functions of fractional fields and their fractional derivatives. Hence we will first derive appropriate fractional Euler–Lagrange equations using an extended variation principle.*

*We will then demonstrate, that the proposed procedure may be applied to the free fields, e.g. for the Dirac matter field leading to the fractional Dirac equation and e.g. for a fractional gauge field, which leads to a fractional extension of the Maxwell equations.*

*In the next chapter we will then extend the principle of local gauge invariance to fractional fields and will indeed determine the exact form of a gauge invariant coupling of the fractional gauge field to the fractional matter field. The result is a fractional gauge invariant extension of the standard QED.*

## 18.1 Fractional Euler–Lagrange equations

The covariant expression for the action $S$ is defined as the time integral of the Lagrangian $L$

$$S = \int L dt \tag{18.6}$$

where the Lagrangian is related to the Lagrangian density $\mathcal{L}$ by

$$L = \int \mathcal{L} d^3x \tag{18.7}$$

As long as the Lagrangian density is a function of the field $\phi$ and its derivatives $\phi_{|\mu}$ the classical Euler–Lagrange-equations follow from the vanishing variation of the action

$$\delta S = 0 \tag{18.8}$$

as

$$\frac{\partial}{\partial \phi}\mathcal{L} - \partial_\mu \frac{\partial \mathcal{L}}{\partial \phi_{|\mu}} = 0 \tag{18.9}$$

We now want to extend this formalism to fractional fields. We will investigate a simple example and derive the fractional Euler–Lagrange-equations for a Lagrangian density, which contains as independent variables the field $\phi$ and its fractional derivatives $\phi^\alpha_{|\mu}$. Starting point is the fractional action:

$$S = \int \mathcal{L}(\phi, \phi^\alpha_{|\mu}) d^3x dt \tag{18.10}$$

We will determine the variation of the action by the limiting procedure

$$\delta S = \lim_{h \to 0} \frac{\Delta S}{h} \tag{18.11}$$

with

$$\Delta S = \int \left( \mathcal{L}(\phi + h\,\delta\phi, \phi^\alpha_{|\mu} + h\,\delta(\phi^\alpha_{|\mu})) - \mathcal{L}(\phi, \phi^\alpha_{|\mu}) \right) d^3x dt \tag{18.12}$$

A series expansion up to first order in $h$ leads to

$$\Delta S = \int \left( h \frac{\partial \mathcal{L}}{\partial \phi} \delta\phi + h \sum_{\mu=0}^{4} \frac{\partial \mathcal{L}}{\partial \phi^\alpha_{|\mu}} \delta\phi^\alpha_{|\mu} \right) d^3x dt \tag{18.13}$$

or

$$\frac{\Delta S}{h} = \int \left( \frac{\partial \mathcal{L}}{\partial \phi} \delta\phi + \sum_{\mu=0}^{4} \frac{\partial \mathcal{L}}{\partial \phi^\alpha_{|\mu}} \delta\phi^\alpha_{|\mu} \right) d^3x dt \tag{18.14}$$

For reasons of simplicity we will ignore surface effects and therefore we demand, that the Lagrangian density vanishes at infinity. In addition we use the property of the scalar product:

$$\int_{-\infty}^{+\infty} f g^\alpha_{|\mu} dx_\mu = \mp \int_{-\infty}^{+\infty} f^\alpha_{|\mu} g dx_\mu \tag{18.15}$$

The minus sign holds for odd fractional derivatives (Liouville-, Fourier-, 3rd, 5th, ... order). The plus sign for all even fractional derivatives (Riesz-, 4th, 6th, ... order).

Hence partial integration of (18.14) leads to:

$$\frac{\Delta S}{h} = \int \left( \frac{\partial \mathcal{L}}{\partial \phi} \delta\phi \mp \sum_{\mu=0}^{4} \partial_\mu^\alpha \left(\frac{\partial \mathcal{L}}{\partial \phi_{|\mu}^\alpha}\right) \delta\phi \right) d^3x dt \tag{18.16}$$

$$= \int \left( \frac{\partial \mathcal{L}}{\partial \phi} \mp \sum_{\mu=0}^{4} \partial_\mu^\alpha \left(\frac{\partial \mathcal{L}}{\partial \phi_{|\mu}^\alpha}\right) \right) \delta\phi \, d^3x dt \tag{18.17}$$

According to (18.8) for $h \to 0$ we obtain the fractional Euler–Lagrange-equations

$$\frac{\partial \mathcal{L}}{\partial \phi} \mp \sum_{\mu=0}^{4} \partial_\mu^\alpha \left( \frac{\partial \mathcal{L}}{\partial \phi_{|\mu}^\alpha} \right) = 0 \tag{18.18}$$

As an example we propose a Lagrangian density for a free fractional Dirac-field:

$$\mathcal{L}_\mathrm{D}^\mathrm{free} = \bar{\Psi}(i\gamma_\alpha^\mu \partial_\mu^\alpha + \mathbf{1}_\alpha m^\alpha)\Psi \tag{18.19}$$

where $\mathbf{1}_\alpha$ indicates a $n \times n$ unit matrix with dimension $n = (2/\alpha)^2$. Using (18.18) the variation with respect to $\bar{\Psi}$ results in the fractional Dirac equation

$$(i\gamma_\alpha^\mu \partial_\mu^\alpha + \mathbf{1}_\alpha m^\alpha)\Psi = 0 \tag{18.20}$$

The hitherto presented strategy for a derivation of Euler–Lagrange-equations may be easily extended to more complex Lagrangian density types (see e.g. [Muslih(2010b)]).

For Lagrangian densities of type

$$\mathcal{L} = \mathcal{L}(\phi, \phi_{|\mu}^\alpha, \phi_{|\mu}^{-\alpha}, \phi_{|\mu}^{-\alpha+1}, ..., \phi_{|\mu}^{-\alpha+k}) \tag{18.21}$$

we obtain a corresponding set of extended fractional Euler–Lagrange-equations:

$$\frac{\partial \mathcal{L}}{\partial \phi} \mp \partial_\mu^\alpha \frac{\partial \mathcal{L}}{\partial (\phi_{|\mu}^\alpha)} \mp \sum_{n=0}^{k} (-1)^n \partial_\mu^{-\alpha+n} \frac{\partial \mathcal{L}}{\partial (\phi_{|\mu}^{-\alpha+n})} = 0 \tag{18.22}$$

## 18.2 The fractional Maxwell equations

In this section we will demonstrate that the direct fractional extension of the ordinary Lagrangian density for the electromagnetic field determines fractional Maxwell equations and consequently leads to fractional wave equations for the electric and magnetic field respectively.

We will first derive these equations for a general fractional derivative and will then discuss a specific use of the Riemann- and Caputo fractional derivative definition in the next section. We define the fractional extension of the standard field strength tensor

$$F^{\alpha}_{\mu\nu} = \partial^{\alpha}_{\nu} A_{\mu} - \partial^{\alpha}_{\mu} A_{\nu} \qquad (18.23)$$

The potential $A_\mu$ is not uniquely determined. Adding a four-gradient of an arbitrary scalar function $\phi^{\alpha}(x, y, z, t)$

$$A^{'}_{\mu} = A_{\mu} + \partial^{\alpha}_{\mu}\phi = A_{\mu} + \phi^{\alpha}_{|\mu} \qquad (18.24)$$

because of

$$\partial^{\alpha}_{\mu}\partial^{\alpha}_{\nu}\phi = \partial^{\alpha}_{\nu}\partial^{\alpha}_{\mu}\phi \qquad (18.25)$$

leaves the form of the field strength tensor unchanged and is therefore called a local gauge transformation.

The Lagrangian density for a fractional electromagnetic field $\mathcal{L}_{\mathrm{fEM}}$ is given by

$$\mathcal{L}_{\mathrm{fEM}} = -\frac{1}{4} F^{\alpha}_{\mu\nu} F^{\mu\nu}_{\alpha} \qquad (18.26)$$

A variation of $\mathcal{L}_{\mathrm{fEM}}$ with respect to the fractional potential $A_\mu$ leads to the fractional inhomogeneous Maxwell equations

$$\partial^{\alpha}_{\mu} F^{\mu\nu}_{\alpha} = 0 \qquad (18.27)$$

since the four-current $j^\nu$ is vanishing in the absence of external sources. We have to show, that the definition of the fractional field strength tensor (18.23) and the corresponding derived field equations (18.27) lead to reasonable results. As a typical example we will derive fractional wave equations which may be used for a description of fractional electromagnetic fields.

Therefore in the following we will derive the vacuum solutions for the fractional Maxwell equations.

We set $\vec{E} = \{E_1, E_2, E_3\}$ and $\vec{B} = \{B_1, B_2, B_3\}$ and present the explicit form for the field strength tensor:

$$F^{\mu\nu}_{\alpha} = \begin{pmatrix} 0 & -E_1 & -E_2 & -E_3 \\ E_1 & 0 & -B_3 & B_2 \\ E_2 & B_3 & 0 & -B_1 \\ E_3 & -B_2 & B_1 & 0 \end{pmatrix} \qquad (18.28)$$

We insert into (18.27):

$$\nabla^\alpha \vec{E} = 0 \tag{18.29}$$
$$\nabla^\alpha \times \vec{B} = \partial_t^\alpha \vec{E} \tag{18.30}$$

where we have introduced the nabla operator $\nabla^\alpha = \{\partial_1^\alpha, \partial_2^\alpha, \partial_3^\alpha\}$.

The homogeneous Maxwell equations may be derived by introducing the dual field strength tensor $\mathcal{F}^{\alpha,\mu\nu}$ which is given with the Levi-Civitta symbol $\epsilon_{\mu\nu\rho\sigma}$ as:

$$\mathcal{F}_{\mu\nu}^\alpha = \frac{1}{2}\epsilon_{\mu\nu\rho\sigma}F^{\alpha,\rho\sigma} \tag{18.31}$$

With the condition

$$\partial_\alpha^\mu \mathcal{F}_{\mu\nu}^\alpha = 0 \tag{18.32}$$

and with

$$F_{\mu\nu}^\alpha = \begin{pmatrix} 0 & E_1 & E_2 & E_3 \\ -E_1 & 0 & -B_3 & B_2 \\ -E_2 & B_3 & 0 & -B_1 \\ -E_3 & -B_2 & B_1 & 0 \end{pmatrix} \tag{18.33}$$

we obtain:

$$\nabla^\alpha \vec{B} = 0 \tag{18.34}$$
$$\nabla^\alpha \times \vec{E} = -\partial_t^\alpha \vec{B} \tag{18.35}$$

The relations (18.29),(18.30),(18.34),(18.35) are the fractional Maxwell equations for the vacuum.

Since we intend to derive a wave equation for the fractional electromagnetic field, we will eliminate $\vec{E}$ and $\vec{B}$ respectively from these equations.

We illustrate the procedure for an elimination of $\vec{E}$:
An application of the curl-operator to (18.30)

$$\nabla^\alpha \times (\nabla_\alpha \times \vec{B}) = \partial_t^\alpha (\nabla_\alpha \times \vec{E}) \tag{18.36}$$

and inserting (18.35) leads to:

$$\nabla^\alpha \times (\nabla_\alpha \times \vec{B}) = -\partial_t^\alpha \partial_\alpha^t \vec{B} \tag{18.37}$$

Let us introduce the fractional Laplace operator $\Delta_\alpha^\alpha$ and the fractional d'Alembert operator $\Box_\alpha^\alpha$.

$$\Delta_\alpha^\alpha = \partial_x^\alpha \partial_\alpha^x + \partial_y^\alpha \partial_\alpha^y + \partial_z^\alpha \partial_\alpha^z \tag{18.38}$$
$$\Box_\alpha^\alpha = \Delta_\alpha^\alpha - \partial_t^\alpha \partial_\alpha^t \tag{18.39}$$

Using the identity $\nabla^\alpha \times (\nabla_\alpha \times \vec{B}) = \nabla^\alpha(\nabla_\alpha \vec{B}) - \Delta^\alpha_\alpha \vec{B}$ and because of $\nabla^\alpha \vec{B} = 0$ we obtain for (18.37):

$$\Delta^\alpha_\alpha \vec{B} = \partial^\alpha_t \partial^t_\alpha \vec{B} \tag{18.40}$$

$$\Box^\alpha_\alpha \vec{B} = 0 \tag{18.41}$$

In a similar procedure we may eliminate $\vec{B}$ from the fractional Maxwell equations and obtain:

$$\Delta^\alpha_\alpha \vec{E} = \partial^\alpha_t \partial^t_\alpha \vec{E} \tag{18.42}$$

$$\Box^\alpha_\alpha \vec{E} = 0 \tag{18.43}$$

Therefore both vectors $\vec{E}$ and $\vec{B}$ fulfil the same type of a fractional wave equation.

Hence we have demonstrated that the use of the fractional Lagrangian density (18.26) leads to a fractional wave equation.

Of course, an interesting question is the physical interpretation and relevance of the proposed fractional Maxwell-equations.

Therefore in the next section we will investigate the properties of fractional interacting fields. We will apply the principle of local gauge invariance to determine the interaction term of a fractional Dirac matter field with a fractional electromagnetic gauge field.

We will obtain a fractional generalization of quantum electrodynamics, which for $\alpha = 1$ reduces to classical QED, which describes with a very high accuracy the interaction of electrons with photons.

Since QED is the most successful theory ever, we will not explore the fractional QED in the vicinity of $\alpha \approx 1$. But for $\alpha \neq 1$ we expect a quantum theory, which extends the standard QED to fractional QED, which describes an interaction of a fractional matter field with a fractional gauge field. Hence the coupling of more complex charge types is described with such a fractional field theory.

We will give a physical interpretation of the proposed fractional fields in terms of quantum chromodynamics (QCD) and will identify the fractional matter field with quarks and the fractional gauge field with gluons.

# Chapter 19

# Gauge Invariance in Fractional Field Theories

*Historically the first example for a quantum field theory is quantum electrodynamics (QED), which successfully describes the interaction between electrons and positrons with photons [Pauli(1941), Sakurai(1967)]. The interaction type of this quantum field theory between the electromagnetic field and the electron Dirac field is determined by the principle of minimal gauge invariant coupling of the electric charge.*

*For more complex charge types, this principle may be extended to non-Abelian gauge theories or Yang–Mills field theories [Yang(1954)]. A typical example is quantum chromodynamics (QCD), which describes a hadronic interaction by gauge invariant coupling of quark fields with gluon fields.*

*Hence local gauge invariance seems to be a fundamental principle for studying of the dynamics of particles [Abers(1973)].*

*In the following we will apply the concept of local gauge invariance to fractional fields and will outline the basic structure of a quantum field theory based on fractional calculus.*

*This concept is motivated by the observation that the fractional Dirac-equation for* $\alpha = 2/3$ *describes particles which obey an inherent* $SU(3)$-*symmetry. Since the only mechanism currently used to introduce a symmetry like* $SU(n)$ *involves non-Abelian gauge fields as proposed by Yang–Mills, the study of gauge invariant fractional wave equations is a new, interesting alternative approach.*

*Free fractional fields have been studied already in the previous chapter [Goldfain(2006), Lim(2006a)]. Different approaches exist on interacting fractional fields [Herrmann(2007), Lim(2008)]. In this chapter we will apply the principle of local gauge invariance to fractional wave equations and derive the resulting interaction up to order* $o(\bar{g})$ *in the coupling constant.*

*The result will be a fractional extension of quantum electrodynamics*

*(QED), which for the limit $\alpha = 1$ reduces to standard QED. But for $\alpha \neq 1$ it describes the interaction of a fractional particle field with a fractional gauge field.*

*As a first application, we will derive the energy spectrum of a non-relativistic fractional particle in a fractional constant magnetic field. The resulting mass formula is lowest-order equivalent to the mass formulas which are previously derived with group theoretical methods and have been successfully applied in the last chapters for an interpretation of meson and baryon spectra.*

## 19.1 Gauge invariance in first order of the coupling constant $\bar{g}$

In the previous chapter we have presented an example for a fractional free field, the relativistic fractional Dirac field and for a fractional gauge field, the fractional electromagnetic field.

We now postulate, that the principle of local gauge invariance is valid for fractional wave equations, too.

The requirement of invariance of the fractional Lagrangian density for the Dirac field under local gauge transformations should uniquely determine the type of the interaction with a fractional gauge field.

We therefore will investigate the transformation properties of the fractional analogue of the free QED-Lagrangian $\mathcal{L}^{\text{free}}_{FQED}$:

$$\mathcal{L}^{\text{free}}_{FQED} = \bar{\Psi}(i\gamma^{\mu}_{\alpha}\partial^{\alpha}_{\mu} + \mathbf{1}_{\alpha}m^{\alpha})\Psi - \frac{1}{4}F^{\beta}_{\mu\nu}F^{\mu\nu}_{\beta} \tag{19.1}$$

where $\alpha = 2/n$ and $\beta$ both denote derivatives of fractional order. $\mathbf{1}_{\alpha}$ indicates a $n \times n$ unit matrix with dimension $n = (2/\alpha)^2$.

Special cases are: for $\alpha = 1$ a fractional electric field couples to the standard Dirac field, for $\beta = 1$ a standard electric field couples to a fractional Dirac field. For $\alpha = \beta$ both fractional fields are of similar type. For $\alpha = \beta = 1$ the Lagrangian density (19.1) reduces to the QED Lagrangian density.

Under local gauge transformations the following transformation properties hold:

$$\bar{\Psi}' = \bar{\Psi}e^{-i\bar{g}\phi(x)} \tag{19.2}$$

$$\Psi' = e^{i\bar{g}\phi(x)}\Psi \tag{19.3}$$

$$A'_{\mu} = A_{\mu} + \phi^{\beta}_{|\mu} \tag{19.4}$$

The transformation properties of the fractional derivative operator $\partial_\mu^\alpha$ may be deduced from:

$$\bar{\Psi}'\gamma_\alpha^\mu\partial_\mu^\alpha\Psi' = \bar{\Psi}'\gamma_\alpha^\mu\partial_\mu^\alpha e^{i\bar{g}\phi}\Psi \tag{19.5}$$

$$= \bar{\Psi}'\gamma_\alpha^\mu(e^{i\bar{g}\phi}\partial_\mu^\alpha + [\partial_\mu^\alpha, e^{i\bar{g}\phi}])\Psi \tag{19.6}$$

$$= \bar{\Psi}\gamma_\alpha^\mu(\partial_\mu^\alpha + e^{-i\bar{g}\phi}[\partial_\mu^\alpha, e^{i\bar{g}\phi}])\Psi \tag{19.7}$$

The commutator reflects the non-locality for the fractional derivative. Since there is no fractional analogue to the chain rule, which is known for the standard derivative, this commutator cannot be evaluated directly.

Therefore we will restrict our derivation to gauge invariance up to first order in the coupling constant $\bar{g}$ only, which corresponds to an infinitesimal gauge transformation.

We obtain:

$$e^{-i\bar{g}\phi}[\partial_\mu^\alpha, e^{i\bar{g}\phi}] = (1 - i\bar{g}\phi)[\partial_\mu^\alpha, 1 + i\bar{g}\phi] \tag{19.8}$$

$$= (1 - i\bar{g}\phi)[\partial_\mu^\alpha, i\bar{g}\phi] \tag{19.9}$$

$$= i\bar{g}[\partial_\mu^\alpha, \phi] + o(\bar{g}^2) \tag{19.10}$$

We define the fractional charge connection operator $\hat{\Gamma}_\mu^{\alpha\beta}$, which in a fractional charge tangent space is the analogue to the fractional Christoffel symbols in a fractional coordinate tangent space or the fractional Fock–Ivanenko coefficients in a fractional spinor tangent space:

$$\hat{\Gamma}_\mu^{\alpha\beta}(\xi_\mu) = \delta_{(\mu)}{}^{\sigma\tau}[\partial_\sigma^\alpha, (\partial_\tau^{-\beta}\xi_\mu)] \tag{19.11}$$

$$= \delta_{(\mu)}{}^{\sigma\tau}[\partial_\sigma^\alpha, \xi_{\mu|\tau}^{-\beta}] \tag{19.12}$$

with the Kronecker-delta $\delta_{(\mu)}{}^{\sigma\tau}$, where $(\mu)$ indicates that a summation over $\mu$ is performed, which reduces the sum (19.11) to a single term (19.12). But on the other hand $(\mu)$ should be ignored, when the transformation properties in fractional coordinate space are determined. Therefore the vector character of $\hat{\Gamma}_\mu^{\alpha\beta}(\xi_\mu)$ is conserved in the fractional coordinate space.

This operator is linear

$$\hat{\Gamma}_\mu^{\alpha\beta}(\xi_\mu + \zeta_\mu) = \hat{\Gamma}_\mu^{\alpha\beta}(\xi_\mu) + \hat{\Gamma}_\mu^{\alpha\beta}(\zeta_\mu) \tag{19.13}$$

It should be emphasized, that $\hat{\Gamma}_\mu^{\alpha\beta}(\xi_\mu)$ is an operator and therefore does not transform as a simple c-number. But up to first order in $\bar{g}$ the relation

$$\bar{\Psi}'\gamma_\alpha^\mu i\bar{g}\hat{\Gamma}_\mu^{\alpha\beta\prime}(\xi_\mu)\Psi' = \bar{\Psi}\gamma_\alpha^\mu i\bar{g}\hat{\Gamma}_\mu^{\alpha\beta}(\xi_\mu)\Psi + o(\bar{g}^2) \tag{19.14}$$

holds. Hence the fractional derivative operator $\partial_\mu^\alpha$ transforms as

$$\bar{\Psi}'\gamma_\alpha^\mu\partial_\mu^\alpha\Psi' = \bar{\Psi}\gamma_\alpha^\mu(\partial_\mu^\alpha + i\bar{g}\hat{\Gamma}_\mu^{\alpha\beta}(\phi_{|\mu}^\beta))\Psi + o(\bar{g}^2) \tag{19.15}$$

Therefore we define the covariant fractional derivative $D_\mu^\alpha$ as

$$D_\mu^\alpha = \partial_\mu^\alpha - i\bar{g}\hat{\Gamma}^{\alpha\beta}_{\ \mu}(A_\mu) \tag{19.16}$$

It transforms as

$$D_\mu^{\alpha'} = \partial_\mu^\alpha + i\bar{g}\hat{\Gamma}^{\alpha\beta}_{\ \mu}(\phi^\beta_{|\mu}) - i\bar{g}\hat{\Gamma}^{\alpha\beta'}_{\ \mu}(A'_\mu) \tag{19.17}$$

With the covariant fractional derivative $D_\mu^\alpha$ it follows from (19.14) and from the linearity of the fractional charge connection operator $\hat{\Gamma}^{\alpha\beta}_\mu$, that the Lagrangian density $\mathcal{L}_{FQED}$ for the fractional extension of QED

$$\mathcal{L}_{FQED} = \bar{\Psi}(i\gamma^\mu_\alpha D^\alpha_\mu + \mathbf{1}_\alpha m^\alpha)\Psi - \frac{1}{4}F^\beta_{\mu\nu}F^{\mu\nu}_\beta \tag{19.18}$$

is invariant under local gauge transformations up to first order in the coupling constant $\bar{g}$.

Variation of $\mathcal{L}_{FQED}$ with respect to $\bar{\Psi}$ leads to the Dirac equation for the fractional Dirac field $\Psi$:

$$(i\gamma^\mu_\alpha \partial^\alpha_\mu + \mathbf{1}_\alpha m^\alpha)\Psi = -\bar{g}\gamma^\mu_\alpha \hat{\Gamma}^{\alpha\beta}_{\ \mu}(A_\mu)\Psi \tag{19.19}$$

Variation of $\mathcal{L}_{FQED}$ with respect to the four-potential $A_\mu$ leads to the fractional Maxwell equations

$$\partial^\beta_\mu F^{\mu\nu}_\beta = -\bar{g}\bar{\Psi}\gamma^\nu_\alpha({\delta_{(\nu)}}^{\sigma\tau}\partial_\tau^{-\beta}\partial_\sigma^\alpha)\Psi \tag{19.20}$$

$$= -\bar{g}\bar{\Psi}\gamma^\nu_\alpha({\delta_{(\nu)}}^{\sigma}\partial_\sigma^{\alpha-\beta})\Psi \tag{19.21}$$

$$= -\bar{g}\bar{\Psi}\gamma^\nu_\alpha \partial_{(\nu)}^{\alpha-\beta}\Psi \tag{19.22}$$

with ${\delta_{(\nu)}}^{\sigma\tau}$ is the Kronecker-delta, where $(\nu)$ indicates, that the term $({\delta_{(\nu)}}^{\sigma\tau}\partial_\tau^{-\beta}\partial_\sigma^\alpha)$ transforms like a scalar in fractional coordinate space but summation is performed in (19.20), which yields a single term $\partial_{(\nu)}^{\alpha-\beta}$ in (19.22), which for $\alpha = \beta$ becomes 1, while for $\alpha > \beta$ fractional differentiation and for $\alpha < \beta$ fractional integration is performed.

The term on the right side of (19.20) is the variation of

$$\mathcal{L}_\Gamma = \bar{g}\bar{\Psi}\gamma^\mu_\alpha\hat{\Gamma}^{\alpha\beta}_{\ \mu}(A_\mu)\Psi \tag{19.23}$$

with respect to the four potential $A_\mu$ and may be derived using the Leibniz product rule for fractional derivatives:

$$\begin{aligned}
\mathcal{L}_\Gamma &= \bar{g}\bar{\Psi}\gamma^\mu_\alpha\hat{\Gamma}^{\alpha\beta}_{\ \mu}(A_\mu)\Psi && (19.24)\\
&= \bar{g}\bar{\Psi}\gamma^\mu_\alpha({\delta_{(\mu)}}^{\sigma\tau}[\partial^\alpha_\sigma, A^{-\beta}_{\mu|\tau}])\Psi\\
&= \bar{g}\bar{\Psi}\gamma^\mu_\alpha({\delta_{(\mu)}}^{\sigma\tau}(\partial^\alpha_\sigma A^{-\beta}_{\mu|\tau} - A^{-\beta}_{\mu|\tau}\partial^\alpha_\sigma))\Psi\\
&= \bar{g}\bar{\Psi}\gamma^\mu_\alpha\left({\delta_{(\mu)}}^{\sigma\tau}\left(\sum_{k=0}^{\infty}\binom{\alpha}{k}A^{-\beta+k}_{\mu|\tau}\partial^{\alpha-k}_\sigma - A^{-\beta}_{\mu|\tau}\partial^\alpha_\sigma\right)\right)\Psi\\
&= \bar{g}\bar{\Psi}\gamma^\mu_\alpha\left({\delta_{(\mu)}}^{\sigma\tau}\sum_{k=1}^{\infty}\binom{\alpha}{k}A^{-\beta+k}_{\mu|\tau}\partial^{\alpha-k}_\sigma\right)\Psi && (19.25)
\end{aligned}$$

where the fractional binomial is given by:

$$\binom{\alpha}{k} = \frac{\Gamma(1+\alpha)}{\Gamma(1+k)\Gamma(1+\alpha-k)} \tag{19.26}$$

Therefore $\mathcal{L}_\Gamma$ is a function of type (18.21). Applying the corresponding Euler–Lagrange equations (18.22) to (19.25) yields

$$\sum_{k=1}^{\infty}(-1)^k \partial_\tau^{-\beta+k} \frac{\partial \mathcal{L}_\Gamma}{\partial (A_{\mu|\tau}^{-\beta+k})} \tag{19.27}$$

$$= \bar{g}\bar{\Psi}\gamma_\alpha^\mu \left( \delta_{(\mu)}{}^{\sigma\tau} \sum_{k=1}^{\infty} \binom{\alpha}{k} (-1)^k \partial_\tau^{-\beta+k} \partial_\sigma^{\alpha-k} \right) \Psi \tag{19.28}$$

$$= \bar{g}\bar{\Psi}\gamma_\alpha^\mu \left( \delta_{(\mu)}{}^{\sigma\tau} \partial_\tau^{-\beta} \partial_\sigma^{\alpha} \sum_{k=1}^{\infty} \binom{\alpha}{k} (-1)^k \right) \Psi \tag{19.29}$$

$$= -\bar{g}\bar{\Psi}\gamma_\alpha^\mu (\delta_{(\mu)}{}^{\sigma\tau} \partial_\tau^{-\beta} \partial_\sigma^{\alpha}) \Psi \tag{19.30}$$

Hence the coupling term in (19.20) is derived.

The non-linear field equations (19.19) and (19.20) completely determine the interaction of the fractional Dirac field with the fractional gauge field up to first order in $o(\bar{g})$. We have proven, that the principle of local gauge invariance still is effective for fractional fields.

Furthermore, besides the mechanism proposed by Yang–Mills, now an alternative approach for implementing an inherent $SU(n)$-symmetry exists. While in a Yang–Mills theory, the $SU(n)$-symmetry is explicitly introduced via a non-Abelian gauge field, now the same symmetry is introduced implicitly via the fractional Dirac equation leaving the symmetry of the fractional gauge field Abelian. Therefore equations (19.18), (19.19) and (19.20) serve as an alternative formulation of a gauge field theory of e.g. for $\alpha = \beta = 2/3$ the strong interaction of hadrons.

For $\alpha = 1, \beta = 1$ these equations reduce to the standard QED equations. For $\alpha \neq 1, \beta \neq 1$ the major extensions are: the electric field $A_\mu$ extends from a c-number to a non-local operator, which we called the fractional charge connection operator. The $\gamma_\alpha^\mu$-matrices obey an extended Clifford algebra, which reflects the transition from $SU(2)$ to $SU(n > 2)$. This makes the fractional QED an interesting candidate to describe properties of particles, which obey an inherent $SU(n)$ symmetry.

To help to get this idea accepted we investigate in the following section the fractional extension of the classical Zeeman effect.

## 19.2 The fractional Riemann–Liouville–Zeeman effect

In the previous section, we have applied a local gauge transformation to a free fractional Dirac field and obtained a fractional interaction with the fractional gauge field. Of course, this method is neither restricted to spinor fields nor does it rely on Lorentz-covariant wave equations, but may be applied to other free fields as well.

In this section, we will present a simple example to illustrate the validity of the proposed mechanism to generate an interaction for fractional fields. We will investigate the fractional analogue of the classical Zeeman effect, which describes the level splitting of a charged particle in an external constant magnetic field.

For that purpose, we will first derive the fractional non-relativistic Schrödinger equation including the interaction term.

The Lagrangian density $\mathcal{L}_{FS}^{\text{free}}$ for the free fractional Schrödinger field is given by

$$\mathcal{L}_{FS}^{\text{free}} = -\frac{1}{2m}(\partial_\alpha^i \Psi)^*(\partial_i^\alpha \Psi) + i\Psi^* \partial_t \Psi \tag{19.31}$$

Local gauge invariance up to first order in $\bar{g}$ is achieved by a replacement of the fractional derivative by

$$D_\mu^\alpha = \{D_t, D_i^\alpha\} = \{\partial_t + i\bar{g}V, \partial_i^\alpha - i\bar{g}\hat{\Gamma}_i^{\alpha\beta}(A_i)\} \tag{19.32}$$

$$= \{D_t, \vec{D}^\alpha\} = \{\partial_t + i\bar{g}V, \nabla^\alpha - i\bar{g}\hat{\vec{\Gamma}}^{\alpha\beta}\} \tag{19.33}$$

Therefore the free Lagrangian density (19.31) is extended to

$$\mathcal{L}_{FS} = -\frac{1}{2m}(D_\alpha^i \Psi)^*(D_i^\alpha \Psi) + i\Psi^*(\partial_t + i\bar{g}V)\Psi \tag{19.34}$$

Variation of $\mathcal{L}_{FS}$ with respect to $\Psi^*$ yields the fractional Schrödinger equation including an interaction with an external gauge field:

$$-\frac{1}{2m}((\partial_i^\alpha - i\bar{g}\hat{\Omega}_i^{\alpha\beta})D_\alpha^i - i(\partial_t + i\bar{g}V))\Psi = 0 \tag{19.35}$$

or

$$-\frac{1}{2m}(\partial_i^\alpha \partial_\alpha^i - i\bar{g}(\hat{\Omega}_i^{\alpha\beta}\partial_\alpha^i + \partial_\alpha^i \hat{\Gamma}_i^{\alpha\beta}))\Psi = i(\partial_t + i\bar{g}V)\Psi \tag{19.36}$$

with

$$\hat{\Omega}_i^{\alpha\beta}(A_i) = -\delta_{(i)}{}^{\sigma\tau} \sum_{k=1}^{\infty} \binom{\alpha}{k} (-1)^k \partial_\sigma^{\alpha-k} A_{i|\tau}^{-\beta+k} \tag{19.37}$$

which results from the variation of $\hat{\Gamma}_i^{\alpha\beta}(A_i)\Psi^*$.

In order to simplify the procedure, we intend to describe a fractional particle in a corresponding fractional constant external magnetic field and set $\alpha = \beta$.

The fractional charge connection operator is then given according to (19.25) by

$$\hat{\Gamma}_i^{\alpha\alpha} = \sum_{k=1}^{\infty} \binom{\alpha}{k} (\partial_i^{-\alpha+k} A_i) \partial_i^{\alpha-k} \tag{19.38}$$

and

$$\hat{\Omega}_i^{\alpha\alpha} = -\sum_{k=1}^{\infty} \binom{\alpha}{k} (-1)^k \partial_i^{\alpha-k} (\partial_i^{-\alpha+k} A_i) \tag{19.39}$$

In order to evaluate these series, we apply the Riemann–Liouville fractional derivative

$$\partial^\alpha x^\nu = \frac{\Gamma(1+\nu)}{\Gamma(1+\nu-\alpha)} x^{\nu-\alpha} \tag{19.40}$$

where $\Gamma(z)$ denotes the Euler $\Gamma$-function.

We now introduce the external field $\vec{A}$:

$$\vec{A} = \{-\frac{1}{2} B y^{2\alpha-1} x^{\alpha-1} z^{\alpha-1}, \frac{1}{2} B x^{2\alpha-1} y^{\alpha-1} z^{\alpha-1}, 0\} \tag{19.41}$$

which for $\alpha = 1$ reduces to $\{-\frac{1}{2}By, \frac{1}{2}Bx, 0\}$ and is therefore the fractional analogue for a constant magnetic field, because

$$\vec{B} = \{B_x, B_y, B_z\} = \nabla^\alpha \times \vec{A} = \{0, 0, B x^{\alpha-1} y^{\alpha-1} z^{\alpha-1}\} \tag{19.42}$$

Since $1/\Gamma(0) = 0$ it follows:

$$\partial_x^\alpha B_z = \partial_y^\alpha B_z = \partial_z^\alpha B_z = 0 \tag{19.43}$$

Therefore the fractional magnetic field is indeed constant within the context of the Riemann–Liouville fractional derivative definition.

Hence we obtain

$$\partial^{-\alpha+1} \vec{A} = \Gamma(\alpha) \{-\frac{1}{2} B y^{2\alpha-1} z^{\alpha-1}, \frac{1}{2} B x^{2\alpha-1} z^{\alpha-1}, 0\} \tag{19.44}$$

and consecutively

$$\partial^k (\partial^{-\alpha+1} \vec{A}) = 0 \tag{19.45}$$

As a consequence, the infinite series in (19.38) and (19.39) reduce to a single term.

$$\begin{aligned} \hat{\vec{\Gamma}}^{\alpha\alpha} &= \Gamma(1+\alpha) \frac{B}{2} \{-z^{\alpha-1} y^{2\alpha-1} \partial_x^{\alpha-1}, z^{\alpha-1} x^{2\alpha-1} \partial_y^{\alpha-1}, 0\} \\ &= \hat{\vec{\Omega}}^{\alpha\alpha} \end{aligned} \tag{19.46}$$

We introduce the fractional analogue of the z-component of the angular momentum operator $\hat{L}_z(\alpha)$

$$\hat{L}_z(\alpha) = i(y^\alpha \partial_x^\alpha - x^\alpha \partial_y^\alpha) \tag{19.47}$$

With this definition we obtain

$$\nabla_\alpha \hat{\vec{\Gamma}}^{\alpha\alpha} = \hat{\vec{\Omega}}^{\alpha\alpha} \nabla_\alpha \tag{19.48}$$

and

$$\nabla_\alpha \hat{\vec{\Gamma}}^{\alpha\alpha} = \hat{\vec{\Omega}}^{\alpha\alpha} \nabla_\alpha = i\Gamma(1+\alpha)\frac{B}{2} z^{\alpha-1} \hat{L}_z(2\alpha - 1) \tag{19.49}$$

The fractional Schrödinger equation (19.35) reduces to

$$-\frac{1}{2m}(\Delta_\alpha^\alpha + \bar{g}\Gamma(1+\alpha)Bz^{\alpha-1}\hat{L}_z(2\alpha-1))\Psi = -i\partial_t \Psi \tag{19.50}$$

where $\Delta_\alpha^\alpha = \nabla^\alpha \nabla_\alpha$. With the product ansatz $\Psi = e^{-iEt}\psi$ a stationary Schrödinger equation for the energy spectrum results in:

$$H\psi = -\frac{1}{2m}(\Delta_\alpha^\alpha + \bar{g}\Gamma(1+\alpha)Bz^{\alpha-1}\hat{L}_z(2\alpha-1))\psi = E\psi \tag{19.51}$$

With (19.51) we have derived the non-relativistic fractional Schrödinger equation for a spinless particle moving in a constant external fractional magnetic field, which is gauge invariant up to first order in $\hat{g}$. We want to emphasize, that for a different choice of the fractional derivative definition a different Hamiltonian results, e.g. for the Caputo fractional derivative definition $z^{\alpha-1}$ has to be replaced by 1.

The derived equation is a fractional integro-differential equation, a solution may be obtained by iteration. In lowest order approximation, we set

$$z^{\alpha-1}\hat{L}_z(2\alpha-1) \approx \rho \hat{L}_z(\alpha) \tag{19.52}$$

with the constant $\rho$ and obtain

$$H^0\psi^0 = -\frac{1}{2m}(\Delta_\alpha^\alpha + \bar{g}\Gamma(1+\alpha)B\rho\hat{L}_z(\alpha))\psi = E_R^0\psi^0 \tag{19.53}$$

Since $\hat{L}_z(\alpha)$ is the Casimir operator of the fractional rotation group $SO^\alpha(2)$, for an analytic solution a group theoretical approach is appropriate.

In order to classify the multiplets only, the Hamiltonian (19.53) may be rewritten as a linear combination of the Casimir operators $\hat{L}^2(\alpha)$ of $SO^\alpha(3)$ which corresponds in a classical picture to the fractional angular momentum and $\hat{L}_z(\alpha)$ of $SO^\alpha(2)$ which corresponds to the z-projection of the angular momentum. The eigenfunctions are determined by two quantum numbers $|LM\rangle$

$$H^0 \,|LM\rangle = m_0 + a_0 \hat{L}^2 + b_0 \hat{L}_z \,|LM\rangle \tag{19.54}$$

The eigenvalues of the Casimir operators based on the Riemann-Liouville fractional derivative definition are given by (12.47) and (12.48)

$$\hat{L}_z(\alpha) \,|LM\rangle = \pm \frac{\Gamma(1+|M|\alpha)}{\Gamma(1+(|M|-1)\alpha)} \,|LM\rangle$$
$$M = 0, \pm 1, \pm 2, ..., \pm L \tag{19.55}$$
$$\hat{L}^2(\alpha) \,|LM\rangle = \frac{\Gamma(1+(L+1)\alpha)}{\Gamma(1+(L-1)\alpha)} \,|LM\rangle$$
$$L = 0, +1, +2, ... \tag{19.56}$$

Note that for $\alpha = 1$ the Casimir operators and the corresponding eigenvalues reduce to the well known results of standard quantum mechanical angular momentum algebra [Edmonds(1957)].

With (19.55) and (19.56) the level spectrum of the multiplets is determined

$$E_R^0 = m_0 + a_0 \frac{\Gamma(1+(L+1)\alpha)}{\Gamma(1+(L-1)\alpha)} \pm b_0 \frac{\Gamma(1+|M|\alpha)}{\Gamma(1+(|M|-1)\alpha)} \tag{19.57}$$

Thus we have derived in lowest order an analytic expression (19.57) for the splitting of the energy levels of a non-relativistic charged fractional spinless particle in a constant fractional magnetic field.

Using this result we may establish a connection with the previously presented results for an interpretation of the hadron spectrum. There we used a heuristic group theoretical approach, which now is justified on the basis of the derived model for the fractional Zeeman-effect.

With reference to (19.51) we may conclude. that a constant fractional magnetic field generates an additional term of the form

$$\frac{1}{2m^{2\alpha-1}} \bar{g} \alpha \Gamma(2-\alpha) B z^{\alpha-1} \hat{L}_z(2\alpha-1) \tag{19.58}$$

According to our derivation presented in Chapter 10 the term $\hat{L}_z(2\alpha-1)$ may be interpreted as a projection of the total angular momentum

$$\hat{L}_z(2\alpha-1) = \hat{L}_z + \hat{S}_z(2\alpha-1) \tag{19.59}$$

and consequently the fractional magnetic field acts on two components simultaneously: the angular momentum and the fractional spin.

These results indicate, that a fractional QED based on the Lagrangian density (19.18) with a small coupling constant $\bar{g}$ and an Abelian gauge field and QCD based on the standard QCD Lagrangian density with a strong coupling constant $\bar{g}$ and a non-Abelian gauge field may probably describe the same phenomena.

Therefore the step to interacting fractional fields based on the principle of local gauge invariance opens up a new exciting research area in high energy and particle physics.

# Chapter 20

# Outlook

In this book we have given a concise introduction to the methods and strategies used within fractional calculus. We have demonstrated, that this specific non-local approach may be a powerful alternative for a description of phenomena in high energy and particle physics. In addition we have shown that the application of fractional group theory leads to a vast amount of intriguing and valuable results.

One reason for this success is the strategy, to interpret concrete experimental data and strictly verify the theoretical results with experimental findings.

Important achievements are:

- Exact prediction of the masses of X(4260) and X(4660) in the charmonium-spectrum, which were verified experimentally afterwards. Prediction of $B_2^*(5818)$ as an excited state of the B-meson, which still is to be verified by experiment.
- The first application of fractional calculus in nuclear theory: a successful description of ground state band excitations in even-even nuclei.
- A derivation of the exact form of interacting fractional fields in lowest order of the coupling constant and a first application: a high-precision mass formula for baryon masses based on the Riemann fractional derivative and a similar mass formula for a simultaneous calculation of all possible mesons based on the Caputo fractional derivative.
- Deduction of the fractional dynamic symmetry group, which simultaneously describes the magic numbers in nuclei and metal clusters.

Furthermore we have outlined, that a unified theory valid for a simultaneous successful description of all fundamental interactions may be of fractional nature.

Many open problems still need to be solved on the way to a consistent formulation of fractional calculus, e.g. a covariant realization of nonlocal operators is a necessary prerequisite.

During the process of writing this book the development of fractional calculus of course keeps evolving, e.g. in this book we restricted the presentation of fractional calculus to a constant fractional derivative parameter $\alpha$. Today we can already estimate, that a generalization of this concept to a derivative of varying order will lead to new insights when applied to systems, where the dynamics is a function of a time and position dependent fractional derivative $\alpha = \alpha(x, t)$ [Samko(1993a), Ramirez(2010)]. This would allow to determine an evolution of the governing differential equations used to describe a complex dynamical system.

This is only one of many examples, where fractional calculus will find its way to establish new strategies to explore previously unknown fields of application. Hence fractional calculus continues to keep ahead in a vividly developing research area. New questions will be raised and new problems will successfully be solved using the fractional approach.

# Bibliography


Abbott, P. C., (2000). *Generalized Laguerre polynomials and quantum mechanics* J. Phys. A: Math. Theor. **33**, 7659–7660.

Abel, N. H., (1823). *Solution de quelques problemes a laide dintegrales definies, Oevres Completes* Vol. **1**, Grondahl, Christiana, Norway, 16–18.

Abers, E. S. and Lee, B. W. (1973). *Gauge theories* Physics Reports C **9**, 1–141.

Abramowitz, M. and Stegun, I. A. (1965). *Handbook of mathematical functions* Dover Publications, New York.

Agrawal, O. P. (2007). *Fractional variational calculus in terms of Riesz fractional derivatives* J. Phys. A: Math. Theor. **40**, 6287–6303.

Amore, P., Fernandez, F. M., Hofmann, C. P. and Saenz, A. (2009). *Collocation method for fractional quantum mechanics* arXiv:0912.2562v1[quant-ph].

Aubert, B. *et al.* (2005). *Observation of a broad structure in the $\pi^+\pi^- J/\Psi$ mass spectrum around 4.26 $GeV/c^2$* Phys. Rev. Lett. **95**, 142001.

Audi, G., Wapstra, A. H. and Thibault, C. (2003). *The AME 2003 atomic mass evaluation* Nucl. Phys. A **729**, 337–676.

Babusci, D., Dattoli, G. and Saccetti, D. (2010). *Integral equations, fractional calculus and shift operator* arXiv:1007.5211v1 [math-ph].

Bacon, R. (1267). *Opus majus* Translated by Robert Belle Burke, Kessinger Publishing LLC, USA (2002).

Baleanu, D. and Muslih, S. (2005). *Lagrangian formulation of classical fields within Riemann Liouville fractional derivatives* Physica Scripta **72**, 119–121.

Baleanu, D. and Trujillo, J. (2010). *A new method of finding the fractional Euler–Lagrange and Hamilton equations within Caputo fractional derivatives* Comm. Nonlin. Sci. Num. Sim. **15**(5), 1111–1115.

Baltay, C. *et al.* (1979). *Confirmation of the existence of the $\Sigma_c^{++}$ and $\Lambda_c^+$ charmed baryons and observation of the decay $\Lambda_c^+ \to \Lambda\pi+$ and $\bar{K}^0 p$* Phys. Rev. Lett. **42**, 1721–1724.

Barnes, T., Close, F. E. and and Swanson, E. S. (1995). *Hybrid and conventional mesons in the flux tube model: Numerical studies and their phenomenological implications* Phys. Rev. D**52**, 5242–5256.

Bender, M., Nazarewicz, W. and Reinhard, P.-G. (2001). *Shell stabilization of super- and hyper-heavy nuclei without magic gaps* Phys. Lett. **B515**, 42–48.

van Beveren, E., Liu, X., Coimbra, R. and Rupp, G. (2009). *Possible $\psi(5S)$, $\psi(4D)$, $\psi(6S)$, and $\psi(5D)$ signals in $\Lambda_c\bar{\Lambda}_c$* Europhys. Lett. **85**, 61002–48.

Bjornholm, S., Borggreen, J., Echt, O., Hansen, K., Pedersen, J. and Rasmussen, H. D. (1990). *Mean-field quantization of several hundred electrons in sodium metal clusters* Phys. Rev. Lett. **65**, 1627–1630.

Bohr, N. and Wheeler, J. A. (1939). *The mechanism of nuclear fission* Phys. Rev. **56**, 426–450.

Bohr, A. (1952). Mat. Fys. Medd. Dan. Vid. Selsk **26**, 14.

Bohr, A. (1954). *Rotational states in atomic nuclei* (Thesis, Copenhagen 1954).

Bollini, C. G. and Giambiagi, J. J. (1993). *Arbitrary powers of d'Alembertian and the Huygens' principle* J. Math. Phys.**34**, 610–621.

Bombelli, R. (1572). *L'Algebra*, Bologna, Italy.

Bonatsos, D. and Daskaloyannis, C. (1999). *Quantum groups and their applications in nuclear physics* Prog. Part. Nucl. Phys. **43**, 537–618.

Bonatsos, D., Lenis, D., Raychev, P. P. and Terziev, P. A. (2002). *Deformed harmonic oscillators in metal clusters: Analytic properties and super-shells* Phys. Rev. A **65**, 033203-033214.

Brack, M. (1993). *The physics of simple metal clusters: self-consistent jellium model and semi-classical approaches* Rev. Mod. Phys. **65**, 677–732.

Burkert, W., (1972). *Lore and science in ancient Pythagoreanism* Harvard University Press, Cambridge (Massachusetts), USA.

Caputo, M. (1967). *Linear model of dissipation whose Q is almost frequency independent Part II* Geophys. J. R. Astr. Soc **13**, 529–539.

Casher, A., Neuberger, H. and Nussinov S. (1979). *Chromoelectric-flux-tube model of particle production* Phys. Rev. D **20**, 179–188.

Casten, R. F., Zamfir, N. V. and Brenner, D. S. (1993). *Universal anharmonic vibrator description of nuclei and critical nuclear phase transitions* Phys. Rev. Lett. **71**, 227–230.

Chowdhury, P. R., Samanta, C. and Basu, D. N. (2006). *$\alpha$ decay half-lives of new super-heavy elements* Phys. Rev. C **73**, 014612.1–014612.6.

Clemenger, K. (1985). *Ellipsoidal shell structure in free-electron metal clusters* Phys. Rev. B **32**, 1359–1362.

Cramer, J. D. and Nix, J. R. (1970). *Exact calculation of the penetrability through two-peaked fission barriers* Phys. Rev. C **2**, 1048–1057.

Davidson, P. M. (1932). *Eigenfunctions for calculating electronic vibrational intensities* Proc. R. Soc. **135**, 459–472.

Descartes, R. (1664). *Le monde ou traité de la lumière* Le Gras, Paris.

Diethelm, K., Ford, N. J., Freed, A. D. and Luchko, Yu. (2005). *Algorithms for the fractional calculus: A selection of numerical methods* Comp. Methods. Appl. Mech. Engrg. **194**, 743–773.

Dirac, P. A. M. (1928). *The quantum theory of the electron* Proc. Roy. Soc. (London) A **117**, 610–624.

Dirac, P. A. M. (1930). *The principles of quantum mechanics* The Clarendon Press, Oxford.

Dirac, P. A. M. (1984). *The requirements of fundamental physical theory* Eur. J. Phys. **5**, 65–67.

Dong, J. and Xu, M. (2007). *Some solutions to the space fractional Schrödinger equation using momentum representation method* J. Math. Phys. **48**, 072105.

Düllmann, Ch. E. *et al.* (2010). *Production and decay of element 114: high cross sections and the new nucleus* $^{277}Hs$ Phys. Rev. Lett. **104**, 252701.

Dürr, S. *et al.* (2008). *Ab initio determination of light hadron masses* Science **322**, 1224–1227.

Eab, C. H. and Lim, S. C. (2006). *Path integral representation of fractional harmonic oscillator* Phys. Lett. A **371**, 303–316.

Edmonds, A. R. (1957). *Angular momentum in quantum mechanics*, Princeton University Press, New Jersey.

Eichten, E., Gottfried, K., Kinoshita, T., Kogut, J., Lane, K, D. and Yan, T, M. (1975). *Spectrum of charmed quark-antiquark bound states* Phys. Rev. Lett. **34**, 369–372 and (1976). *Interplay of confinement and decay in the spectrum of charmonium* Phys. Rev. Lett. **36**, 500–504.

Einstein, A., Podolsky, B. and Rosen, N. (1935). *Can quantum-mechanical description of physical reality be considered complete?*, Phys. Rev. **47**, 777–780.

Eisenberg, J. M. and Greiner, W. (1987). *Nuclear models* North Holland, Amsterdam.

Elsasser, W. M. (1933). *Sur le principe de Pauli dans les noyaux* J. Phys. Radium **4**, 549–556, *ibid.* **5** (1934), 389–397, *ibid.* **5** (1934), 635–639.

Elliott, J. P. (1958). *Collective motion in the nuclear shell model. I. Classification schemes for states of mixed configurations* Proc. Roy. Soc. London A **245**, 128–145.

Erzan, A. and Eckmann, J.-P. (1997). *q-analysis of fractal sets* Phys. Rev. Lett. **78**, 3245–3248.

Euler, L. (1738). *De progressionibus transcendentibus seu quarum termini generales algebraice dari nequeunt* Commentarii academiae scientiarum Petropolitanae **5**, 36–57. Reprinted in Opera omnia I.14, 1–24.

Feller, W. (1952). *On a generalization of Marcel Riesz' potentials and the semi-groups generated by them* Comm. Sem. Mathem. Universite de Lund, 72–81.

Feng, D. H., Gilmore, R. and Deans, S. R. (1981). *Phase transitions and the geometric properties of the interacting boson model* Phys. Rev. C **23**, 1254–1258.

Fermi, E. (1934). *Possible production of elements of atomic number higher than 92* Nature **133**, 898–899.

Feynman, R. P. (1949). *The theory of positrons* Phys. Rev. **76**, 749–759.

Fibonacci (1204). *Liber Abaci*, translated by Sigler, L., Springer, New York (2002).

Fink, H. J., Maruhn, J., Scheid, W. and Greiner, W. (1974). *Theory of fragmentation dynamics in nucleus-nucleus collisions* Z. Physik **268**, 321–331.

Flerov, G. N. and Petrzhak, K. A. (1940). *Spontaneous fission of uranium* J. Phys. **3**, 275–280.

Fortunato, L. (2005). *Solutions of the Bohr Hamiltonian, a compendium* Eur. Phys. J. A **26** s01, 1–30.

Fourier, J. B. J. (1822). *Thêorie analytique de la chaleur* Cambridge University press (2009), Cambridge, UK.

Frank, A., Jolie, J. and Van Isacker, P. (2009). *Symmetries in atomic nuclei* Springer, Berlin, Heidelberg, New York.

Fridmann, J., Wiedenhöver, I., Gade, A., Baby, L. T., Bazin, D., Brown, B. A., Campbell, C. M., Cook, J. M., Cottle, P. D., Diffenderfer, E., Dinca, D.-C., Glasmacher, T., Hansen, P. G., Kemper, K. W., Lecouey, J. L., Mueller, W. F., Olliver, H., Rodriguez-Vieitez, E., Terry, J. R., Tostevin, J. A. and Yoneda, K. (2005). *Magic nucleus* $^{42}Si$ Nature **435**, 922–924.

Gauß, C. F. (1799). *Demonstratio nova theorematis omnem functionem algebraicam rationalem integram unius variabilis in factores reales primi vel secundi gradus resolvi posse* dissertation, apud Fleckeisen, C. G., Helmstedt, Germany.

Gneuss, G. and Greiner, W. (1971). *Collective potential energy surfaces and nuclear structure* Nucl. Phys. A **171**, 449–479.

Goodfrey, S. and Isgur, N. (1985). *Mesons in a relativized quark model with chromodynamics* Phys. Rev. D **32**, 189–231.

Goeppert-Mayer, M. (1949). *On closed shells in nuclei* Phys. Rev. **75**, 1969-1970.

Goldfain, E., (2006). *Complexity in quantum field theory and physics beyond the standard model* Chaos, Solitons and Fractals **28**, 913–922.

Goldfain, E., (2008a). *Fractional dynamics and the TeV regime of field theory* Communications in non linear science and numerical simulation **13**, 666–676.

Goldfain, E., (2008b). *Fractional dynamics and the standard model for particle physics* Communications in non linear science and numerical simulation **13**, 1397–1404.

Gorenflo, R. and Mainardi, F. (1996). *Fractional oscillations and Mittag-Leffler functions* in: Proceedings of RAAM 1996, Kuwait University, 193–196.

Gorenflo, R. and Mainardi, F. (1997). *Fractional calculus: integral and differential equations of fractional order* in: Carpinteri, A. and Mainardi, F. :(editors): *Fractals and Fractional Calculus in Continuous Mechanics* Springer Verlag, Wien and New York, 223-276.

Gorenflo, R. *et al.* (2008). `http:\\www.fracalmo.org`.

Greiner, M., Scheid, W. and Herrmann, R. (1988). *Collective spin by linearization of the Schrödinger equation for nuclear collective motion* Mod. Phys. Lett. A **3(9)**, 859–866.

Greiner, W. and Müller, B. (1994). *Quantum mechanics - symmetries* Springer, Berlin, Heidelberg, New York.

Greiner, W., Park, J. A. and Scheid, W. (1995). *Nuclear molecules* World Scientific, Singapore.

Greiner, W. and Neise, L. (2001). *Thermodynamics and statistics* Springer Berlin, New York.

Grünwald, A. K. (1867). *Über begrenzte Derivationen und deren Anwendung* Z. angew. Math. und Physik **12**, 441–480.

Guo, X and Xu, M. (2006). *Some physical applications of fractional Schrödinger equation* J. Math. Phys. **47**, 082104–082113.

Gupta, R. K., Cseh, J., Ludu, A., Greiner, W. and Scheid, W. (1992). *Dynamical symmetry breaking in $SU(2)$ model and the quantum group $SU(2)_q$* J. Phys. G: Nucl. Part. Phys. **18**, L73–L82.

Haapakoski, P., Honkaranta, P. and Lipas, P. O. (1970). *Projection model for ground bands of even-even nuclei* Phys. Lett. B **31**, 493–495.

Hahn, O. and Straßmann, F. (1939). *Über den Nachweis und das Verhalten der bei der Bestrahlung des Urans mittels Neutronen entstehenden Erdalkalimetalle* Die Naturwissenschaften **27**, 11–15.

Hall, A. R. (1952). *Ballistics in the seventeenth century* Cambridge University Press, Cambridge.

Han, T. (2010). *The dawn of the LHC era* World Scientific, Singapore.

Haxel, F. P., Jensen, J. H. D. and Suess, H. D. (1949). *On the magic numbers in nuclei* Phys. Rev. **75**, 1766-1766.

Heer, W. A. de (1993). *The physics of simple metal clusters: experimental aspects and simple models* Rev. Mod. Phys. **65**(3), 611–676.

Heisenberg, W. (1927). *Über den anschaulichen Inhalt der quantentheoretischen Kinematik und Mechanik* Zeitschrift f. Physik **43**, 172–198.

Henriquez, A. B. (1979). *A study of charmonium, upsilon and strangeonium systems* Z. Phys. C **2**, 309–312.

Herrmann, R., Plunien, G., Greiner, M., Greiner, W. and Scheid, W. (1989). *Collective spin from the linearization of the Schrödinger equation in multidimensional Riemannian spaces used in collective nuclear models* Int. J. Mod. Phys. A **4**(18), 4961–4975.

Herrmann, R. (2005). *Continuous differential operators and a new interpretation of the charmonium spectrum* arXiv:nucl-th/0508033. *Properties of a fractional derivative Schrödinger type wave equation and a new interpretation of the charmonium spectrum* arXiv:math-ph/0510099.

Herrmann, R. (2006). *The fractional symmetric rigid rotor* arXiv:nucl-th/0610091v1, J. Phys. G: Nucl. Part. Phys. **34**, 607–625.

Herrmann, R. (2007). *Gauge invariance in fractional field theories* arXiv:0708.2262v1 [math-ph], Phys. Lett. A **372**(34), (2008) 5515–5522.

Herrmann, R. (2008a). *Riemann curvature in collective spaces* arXiv:0801.0298 [nucl-th].

Herrmann, R. (2008b). *Higher-dimensional mixed fractional rotation groups as a basis for dynamic symmetries generating the spectrum of the deformed Nilsson oscillator* arXiv:0806.2300v1 [physics.gen-ph], Physica A **389**(4), (2010) 693–704.

Herrmann, R. (2009a). *Higher order fractional derivatives* arXiv:0906.2185v2 [physics.gen-ph].

Herrmann, R. (2009b). arXiv:0907.1953v1 [physics.gen-ph], *Fractional phase transition in medium size metal clusters* Physica A **389**(16), (2010) 3307–3315.

Herrmann, R. (2010a). *Fractional quantum numbers deduced from experimental ground state meson spectra* arXiv:1003.5246v1 [physics.gen-ph].

Herrmann, R. (2010b). *Common aspects of q-deformed Lie-algebras and fractional calculus* arXiv:0711.3701v1 [physics.gen-ph], arXiv:1007.1084v1 [physics.gen-ph], Physica A **389**(21), (2010) 4613–4622.

Higgs, P. W. (1964). *Broken symmetries and the masses of gauge bosons* Phys. Rev. Lett. **13**, 508-509.

Hilfer, R. (2000). *Applications of fractional calculus in physics* World Scientific Publishing, River Edge, NJ.

Hilfer, R. and Seybold, H. J. (2006). *Computation of the generalized Mittag-Leffler function and its inverse in the complex plane* Integral transforms and special functions **17**(9), 637–652.

Hilfer, R. (2008). *Threefold introduction to fractional derivatives* in: Klages, R., Radons, G. and Sokolov I. (editors). *Anomalous transport, foundations and application* Wiley-VCH, Weinheim, Germany, 17–73.

Hofmann, S., Ninov, V., Heßberger, F. P., Armbruster, P., Folger, H., Münzenberg, G., Schött, H. J., Popeko, A. G., Yeremin, A. V., Saro, S., Janik, R. and Leino, M. (1996). *The new element 112* Z. Phys. A **354**, 229–230.

Hofmann, S. and Münzenberg, G. (2000). *The discovery of the heaviest elements* Rev. Mod. Phys. **72**, 733–767.

Hu, Y and Kallianpur, G. (2000). *Schrödinger equations with fractional Laplacians* Appl. Math. Optim. **42**, 281–290.

Hullmeine, U., Winsel, A. and Voss E. (1989). *Effect of previous charge/discharge history on the capacity of the $PbO_2/PbSO_4$ electrode: the hysteresis or memory effect* J. of Power Sources **25**(1), 27–47.

Hurwitz, A. (1882). *Einige Eigenschaften der Dirichlet'schen Funktionen $F(s) = \sum(\frac{D}{n})\frac{1}{n^s}$, die bei der Bestimmung der Klassenanzahlen binärer quadratischer Formen auftreten* Z. für Math. und Physik **389**(27), 86-101.

Iachello, F. and Arima, A. (1987). *The interacting boson model* Cambridge University Press, Cambridge.

Iomin, A. (2009). *Fractional time quantum mechanics* Phys. Rev. E **80**(2), 022103.

Jeng, M., Xu, S.-L.-Y., Hawkins, E. and Schearz J. M. (2008). *On the nonlocality of the fractional Schrödinger equation* arXiv:0810.1543v1 [math-ph], American Physical Society, 2009 APS March Meeting, March 16-20, 2009, abstract no.: Q14.013.

Kant, I. (1781). *Anmerkung zur dritten Antinomie* in: Die Antinomie der reinen Vernunft, Akademie Verlag, Berlin 1998.

Kalugampola, U. (2010). *New approach to a generalized fractional integral* arXiv:1010.0742v1 [math.CA]

Kerner, R. (1992). *$Z_3$-grading and the cubic root of the Dirac equation* Classical Quantum Gravity **9**, S137–S146.

Kilbas, A. A., Srivastava, H. M. and Trujillo, J. J. (2003). *Fractional differential equations:A emergent field in applied and mathematical sciences* in Samko, S., Lebre, A. and Dos Santos, A. F. (Eds.) (2003). *Factorization, singular operators and related problems, Proceedings of the conference in honour of professor Georgii Litvinchuk* pp. 151–175, Springer Berlin, Heidelberg, New York.

Kilbas, A. A., Srivastava, H. M. and Trujillo, J. J. (2006). *Theory and applications of fractional differential equations* Elsevier, Amsterdam.

Kiryakova, V. S. (1994). *Generalized fractional calculus and applications* Longman (Pitman Res. Notes in Math. Ser. **301**), Harlow; co-publ.: J. Wiley and Sons, New York.

Knight, W. D., Clemenger, K., de Heer, W. A., Saunders, W. A., Chou, M. Y. and Cohen, M. L. (1984). *Electronic shell structure and abundances of sodium clusters* Phys. Rev. Lett. **52**, 2141–2143.

Korteweg, D. J. and de Vries, G. (1895). *On the change of form of long waves advancing in a rectangular canal, and on a new type of long stationary waves* Philosophical Magazine **39**, 422–443.

Krammer, M. and Krasemann, H. (1979). *Quarkonia in quarks and leptons* Acta Physica Austriaca Suppl. XXI, 259–266.

Kruppa, A. T., Bender, M., Nazarewicz, W., Reinhard, P. G., Vertse, T. and Cwiok, S. (2000). *Shell corrections of superheavy nuclei in self-consistent calculations* Phys. Rev. C **61**, 034313–034325.

Lämmerzahl, C. (1993). *The pseudo-differential operator square root of the Klein-Gordon equation* J. Math. Phys. **34**, 3918–3932.

Laskin, N. (2000). *Fractals and quantum mechanics* Chaos **10**, 780–791.

Laskin, N. (2002). *Fractional Schrödinger equation* Phys. Rev. E **66**, 056108–0561014.

Laskin, N. (2010). *Principles of fractional quantum mechanics* arXiv:1009.5533v1 [math.ph].

Lee, T. D., Oehme, R. and Yang, C. N. (1957). *Remarks on possible noninvariance under time reversal and charge conjugation* Phys. Rev. **106**, 340–345.

Leibniz, G. F. (1675). *Methodi tangentium inversae exempla*, manuscript.

Leibniz G F (1695). *Correspondence with l'Hospital*, manuscript.

Levy-Leblond, J. M. (1967). *Nonrelativistic particles and wave equations* Comm. Math. Phys. **6**, 286–311.

Li, B.-Q. and Chao, K. T. (2009). *Higher charmonia and X,Y,Z states with screened potential* Phys. Rev. D**79**, 094004–094017.

Li, M.-F., Ren J.-R. and Zhu, T. (2010). *Fractional vector calculus and fractional special function* arXiv:1001.2889v1[math.ph].

Lighthill, M. J. (1958). *Introduction to Fourier analysis and generalized functions* Cambridge University Press, Cambridge.

Lim, S. C. and Muniandy, S. V. (2004). *Stochastic quantization of nonlocal fields* Phys. Lett. A **324**, 396–405.

Lim, S. C. (2006a). *Fractional derivative quantum fields at finite temperatures* Physica A **363**, 269–281.

Lim, S. C. and Eab, C. H. (2006b). *Riemann–Liouville and Weyl fractional oscillator processes* Phys. Lett. A **335**, 87–93.

Lim, S. C. and Tao, L. P. (2008). *Topological symmetry breaking of self-interacting fractional Klein–Gordon field theories on toroidal spacetime* J. Phys. A: Math. Theor. **41**, 145403–145432.

Lim, S. C. and Tao, L. P. (2009). *Repulsive Casimir force from fractional Neumann boundary conditions* Phys. Lett. B **679**, 130–137.

Liouville, J. (1832). *Sur le calcul des differentielles á indices quelconques* J. École Polytechnique **13**, 1–162.

Louck, J. D. and Galbraith, H. W. (1972). *Application of orthogonal and unitary group methods to the N-body problem* Rev. Mod. Phys. **44**(3), 540–601.

Lustig, H.-J., Maruhn, J. A. and Greiner, W. (1980). *Transitions in the fission mass distributions of the fermium isotopes* J. Phys. G **6**, L25–L36.

Mainardi, F. (1996). *Fractional relaxation-oscillation and fractional diffusion-wave phenomena* Chaos, Solitons and Fractals **7**(9), 1461–1477.

Mainardi, F. and Gorenflo, R. (2000). *On Mittag-Leffler-type functions in fractional evolution processes* J. Comput. Appl. Math. **118**(1-2), 283–299.

Mainardi, F. (2010). *Fractional calculus and waves in linear viscoelasticity: An introduction to mathematical models* World Scientific, Singapore.

Mandelbrot, B. (1982). *Fractal geometry of nature* W. H. Freeman, New York.

Marchant, T. R. and Smyth, N. F. (1990). *The extended Korteweg–de Vries equation and the resonant flow of a fluid over topology* J. Fluid Mech. **221**, 263–288.

Martin, T. P., Bergmann, T., Göhlich, H. and Lange, T. (1991). *Electronic shells and shells of atoms in metallic clusters* Z. Phys. D **19**, 25–29.

Martin, R. R. (2006). *Nuclear and particle physics* John Wiley and Sons, Hoboken, NJ, USA.

Maruhn, J. A. and Greiner, W. (1972). *The asymmetric two center shell model* Z. Physik **251**, 431–457.

Maruhn, J. A., Hahn, J., Lustig, H.-J., Ziegenhain, K.-H. and Greiner, W. (1980). *Quantum fluctuations within the fragmentation theory* Prog. Part. Nucl. Phys. **4**, 257–271.

May, R. M. (1975). *Stability and complexity in model ecosystems* Princeton Univ. Press, Princeton, second edition.

Meijer, C. S. (1941). *Multiplikationstheoreme für die Funktion $G^{m,n}_{p,q}(z)$*. Proc. Niederl. Akad. Wetenschau **44**, 1042–1070.

Meitner, L. and Fritsch, O. R. (1939). *Disintegration of uranium by neutrons: A new type of nuclear reaction* Nature **143**, 239–240.

Messiah, A. (1968). *Quantum mechanics* John Wiley & Sons, North-Holland Pub. Co, New York.

Mie, G. (1908). *Beiträge zur Optik trüber Medien, speziell kolloidaler Metallösungen* Ann. d. Physik **25**(3), 377–379.

Miller, K. and Ross, B. (1993). *An introduction to fractional calculus and fractional differential equations* Wiley, New York.

Mittag-Leffler, M. G. (1903). *Sur la nouvelle function $E_\alpha(x)$* Comptes Rendus Acad. Sci. Paris **137**, 554–558.

Möller P., Nix, J. R., Myers, W. D. and Swiatecki, W. J. (1995). *Nuclear ground-state masses and deformations* Atomic Data Nucl. Data Tables **59**(2), 185–381.

Möller, P., Bengtsson, R., Carlsson, B. G. Olivius, P., Ichikawa, T. Sagawa, H. and Iwamoto, A. (2008). *Axial and reflection asymmetry of the nuclear ground state* Atomic Data and Nuclear Data Tables **94**(5), 758–780.

Moriarty, P. (2001). *Nanostructured materials* Rep. Prog. Phys. **64**, 297–381.

Mosel, U. and Greiner, W. (1969). *On the stability of superheavy nuclei against fission* Z. Phys. A **222**, 261–282.

Münster, S. (1551). *Rudimenta mathematica* Petri, Basel, **1**, page 29. digital edition: *Sächsische Landesbibliothek - Staats- und Universitätsbibliothek Dresden*, Math.59, misc.2, `http://digital.slub-dresden.de/ppn274384116`.

Muslih, S. I., Agrawal, O. P. and Baleanu, D. (2010a). *A fractional Dirac equation and its solution* J. Phys. A: Math. Theor. **43**(5), 055203.

Muslih, S. I. (2010b). *A formulation of Noethers theorem for fractional classical fields* arXiv:1003.0653v1 [math-ph].

Myers, W. D. and Swiatecki, W. J. (1966). *Nuclear masses and deformations* Nucl. Phys. **81**, 1–60.

Naber, M. (2004). *Time fractional Schrödinger equation* J. Math. Phys. **45**, 3339–3352.

El-Nabulsi, R. A. and Torres, D. F. M. (2008). *Fractional actionlike variational problems* J. Math. Phys. **49**(5) 053521–053527.

El-Nabulsi, R. A. (2009). *Complexified quantum field theory and mass without mass from multidimensional fractional actionlike variational approach with dynamical fractional exponents* Chaos, Solitons and Fractals, **42** (4), 2384–2398.

Nakamura, K. *et al.* (2010). (Particle Data Group) *Review of particle physics* J. Phys. G: Nucl. Part. Phys. **37**, 075021. (URL: http://pdg.lbl.gov).

Newton, I. (1669). *De analysi per aequitiones numero terminorum infinitas*, manuscript.

Nishioka, H., Hansen, K. and Mottelson, B. R. (1990) *Supershells in metal clusters* Phys. Rev. B **42**(15), 9377–9386.

Nigmatullin, R. R. (1992) *Fractional integral and its physical interpretation* Teoret. Mat. Fiz. **90**(3), 354–368.

Nilsson, S. G. (1955). Kgl. Danske Videnskab. Selsk. Mat.-Fys. Medd. **29**, 431.

Nilsson, S. G., Tsang, C. F., Sobiczewski, A., Szyma, Z., Wycech, S., Gustafson, C., Lamm, I., Möller, P. and Nilsson, B. (1969). *On the nuclear structure and stability of heavy and super-heavy elements* Nucl. Phys. A **131**, 1–66.

Niyti, Gupta, R. K. and Greiner, W. (2010) *Establishing the island of stability for superheavy nuclei via the dynamical cluster-decay model applied to a hot fusion reaction* $^{48}Ca + {}^{238}U \rightarrow {}^{286}112$ J. Phys. G: Nucl. Part. Phys. **37**, 115103.

Noddack, I. (1934). *Über das Element 93* Angewandte Chemie **47**, 653–655.

Noether, E. (1918). *Invarianten beliebiger Differentialausdrücke* Gött. Nachr. **1918**, 37–44, *Invariante Variationsprobleme ibid.*, 235–257.

Nogin, V. A., Samko, S. G. (1999). *Some applications of potentials and approximative inverse operators in multi-dimensional fractional calculus* Fract. Calculus and Appl. Anal. **2**, 205–228.

Nomura, K., Shimizu, N. and Otsuka, T. (2008). *Mean-field derivation of the interacting boson model Hamiltonian and exotic nuclei* Phys. Rev. Lett. **101**, 142501.1–142501.4.

Oganessian, Yu. Ts., Utyonkov, V. K., Lobanov, Yu. V., Abdullin, F. Sh., Polyakov, A. N., Shirokovsky, I. V., Tsyganov, Yu. S., Gulbekian, G. G., Bogomolov, S. L., Gikal, B. N., Mezentsev, A. N., Iliev, S., Subbotin, V. G., Sukhov, A. M., Voinov, A. A., Buklanov, G. V., Subotic, K., Zagrebaev, V. I. and Itkis, M. G. (2004). *Measurements of cross sections for the fusion-evaporation Reactions Pu-244(Ca-48,xn)(292-x)114 and Cm-245(Ca-48,xn)(293-x)116* Phys. Rev. C **69**, 054607–0546015.

Oganessian, Yu. Ts., Abdullin, F. Sh., Bailey, P. D., Benker, D. E., Bennett, M. E., Dmitriev, S. N., Ezold, J. G., Hamilton, J. H., Henderson, R. A., Itkis, M. G., Lobanov, Yu. V., Mezentsev, A. N., Moody, K. J., Nelson, S. L., Polyakov, A. N., Porter, C. E., Ramayya, A. V., Riley, F. D., Roberto, J. B., Ryabinin, M. A., Rykaczewski, K. P., Sagaidak, R. N., Shaughnessy, D. A., Shirokovsky, I. V., Stoyer, M. A., Subbotin, V. G., Sudowe, R., Sukhov, A. M., Tsyganov, Yu. S., Utyonkov, V. K., Voinov, A. A., Vostokin, G. K. and Wilk, P. A. (2010). *Synthesis of a new element with atomic number* $Z = 117$ Phys. Rev. Lett. **104**, 142502.1–142502.4.

Oldham, K. B. and Spanier, J. (1974). *The fractional calculus*, Academic Press, New York.

Ortiguera, M. D. (2003). *On the initial conditions in continuous-time fractional linear systems* Sign. Proc. **83**, 2301–2309.

Pakhlova, G. V., Pakhlov, P. N. and Eidel'man, S. I. (2010). *Exotic charmonium* Uspekhi Fizicheskikh Nauk **180**(3), 225–248.

Pauli, W. (1941). *Relativistic field theories of elementary particles* Rev. Mod. Phys. **13**, 203–232.

Plessis, N. du (1957). *Spherical fractional integrals* Trans. Amer. Math. Soc. **84** (1), 262–272.

Plyushchay, M. S. and Rausch de Traubenberg, M. (2000). *Cubic root of Klein-Gordon equation* Phys. Lett. B **477**(1-3), 276–284.

Podlubny, I. (1999). *Fractional differential equations*, Academic Press, New York.

Podlubny, I. (2002). *Geometric and physical interpretation of fractional integration and fractional differentiation* Fractional Calculus and Applied Analysis, **5**(4), 367–386.

Poenaru, D. N., Gherghescu, R. A. and Greiner, W. (2010). *Individual and collective properties of fermions in nuclear and atomic cluster systems* J. Phys. G: Nucl. Part. Phys. **37**, 085101–085110.

Prudnikov, A. P., Brychkov, Yu. A. and Marichev, O. I. (1990). *Integrals and Series, Volume 3: More special functions* Gordon and Breach Science Publishers, New York.

Prusinkiewicz, P. and Lindenmayer, A. (1993). *The algorithmic beauty of plants* Springer, Berlin, Heidelberg, New York.

Racah, G. (1943). *Theory of complex spectra* Phys. Rev. **63**, 367–382.

Raduta, A. A., Gheorghe, A. C. and Faessler, A. (2005). *Remarks on the shape transition from spherical to deformed gamma unstable nuclei* J. Phys. G: Nucl. Part. Phys. **31**, 337–353.

Raduta, A. A., Budaca, R. and Faessler, A. (2010). *Closed formulas for ground band energies of nuclei with various symmetries* J. Phys. G: Nucl. Part. Phys. **37**, 085108–085133.

Ramirez, L. E. S. and Coimbra, C. F. M. (2010). *On the selection and meaning of variable order operators for dynamic modelling* Int. J. Differential Equations **2010**, Article ID 846107.

Raspini, A. (2000). *Dirac equation with fractional derivatives of order 2/3* Fizika B **9**, 49–55.

Raspini, A. (2001). *Simple solutions of the fractional Dirac equation of order 2/3* Physica Scripta **64**(1), 20–22.

Raychev, P. P, Roussev, R. P. and Smirnov, Yu. F. (1990). *The quantum algebra $SU_q(2)$ and rotational spectra of deformed nuclei* J. Phys. G: Nucl. Part. Phys. **16**, L137–L142.

Riemann, B. (1847). *Versuch einer allgemeinen Auffassung der Integration und Differentiation* in: Weber H (Ed.), *Bernhard Riemann's gesammelte mathematische Werke und wissenschaftlicher Nachlass*, Dover Publications (1953), 353.

Riesz, M. (1949). *L'integrale de Riemann-Liouville et le probléme de Cauchy* Acta Math. **81**, 1–223.

Roberts, M. D. (2009). *Fractional derivative cosmology* arXiv:0909.1171v1.

Rose, M. E. (1995). *Elementary theory of angular momentum*, Dover Publications, New York.

Ross, B. (1977). *Fractional calculus* Math. Mag. **50**(3), 115–122.

Rowling, J. K. (2000). *Harry Potter and the prisoner of Azkaban* Bloomsbury Publishing, UK.

Rubin, B. (1996). *Fractional integrals and potentials*, Pitman Monographs and Surveys in Pure and Applied Mathematics, vol. 82, Longman, Harlow.

Rufa, M., Reinhard, P. G., Maruhn, J. A., Greiner, W. and Strayer, M. R. (1988). *Optimal parametrization for the relativistic mean-field model of the nucleus* Phys. Rev. C **38**, 390–409.

Rutz, K., Bender., M., Bürvenich, T., Schilling, T., Reinhard, P. G., Maruhn, J. A. and Greiner, W. (1997). *Superheavy nuclei in self-consistent nuclear calculations* Phys. Rev. C **56**, 238-243.

Ryabov, Ya. E. and Punzenko, A. (2002). *Damped oscillations in view of the fractional oscillator equation* Phys. Rev. B **66**, 184201–184208.

Sakurai, J. J. (1967). *Advanced quantum mechanics* Benjamin/Cummings, New York.

Samko, S. G. and Ross, B. (1993). *Integration and differentiation to a variable fractional order* Integral Transforms and Special Functions **1**(4), 277–300.

Samko, S. G., Kilbas, A. A. and Marichev, O. I. (1993). *Fractional integrals and derivatives* Translated from the 1987 Russian original, Gordon and Breach, Yverdon.

Sandev, T. and Tomovski, Z. (2010). *The general time fractional wave equation for a vibrating string* J. Phys. A: Math. Theor. **43**, 055204.

Sandulescu, A., Gupta, R. K., Scheid, W. and Greiner W. (1976). *Synthesis of new elements within the fragmentation theory: Application to Z = 104 and 106 elements* Phys. Lett. B **60**(3), 225–228.

Sato, Y., Takeuchi, S. and Kobayakawa K. (2001). *Cause of the memory effect observed in alkaline secondary batteries using nickel electrode* J. of Power Sources **93**(1-2), 20–24.

Scharnweber, D., Mosel, U. and Greiner, W. (1970). *Asymptotically correct shell model for nuclear fission* Phys. Rev. Lett. **24**, 601–603.

Schmutzer, E. (1968). *Relativistische Physik* Teubner, Leipzig.

Segovia, J., Entem, D. R. and Fernandez, F. (2010). *Charmonium narrow resonances in the string breaking region* J. Phys. G: Nucl. Part. Phys. **37**, 075010–075021.

Serot, B. D. and Walecka, J. D. (1986). *The relativistic nuclear many-body problem* Adv. Nucl. Phys. **16**, 1–327.

Seybold, H. J. and Hilfer, R. (2005). *Numerical results for the generalized Mittag-Leffler function* FCAA **8**(2),127–139.

Seybold, H. J. and Hilfer, R. (2008). *Numerical algorithm for calculating the generalized Mittag-Leffler function* Siam J. Numer. Anal. **47**(1), 69–88.

Sobiczewski, A. and Pomorski, K. (2007). *Description of structure and properties of super-heavy nuclei* Prog. Part. Nucl. Phys. **58**, 292–349.

Spanier, J., Myland, J. and Oldham, K. B. (2008). *An atlas of functions* Springer, Heidelberg, New York.

Stancu, F. I. (2010). *Can Y(4140) be a $c\bar{c}s\bar{s}$ tetraquark?* J. Phys. G: Nucl. Part. Phys. **37**, 075017.

Stanislavsky, A. A. (2000). *Memory effects and macroscopic manifestation of randomness* Phys. Rev. E **61**, 4752–4759.

Stanislavsky, A. A. (2004). *Fractional oscillator* Phys. Rev. E **70**, 051103–051109.

Stanislavsky, A. A. (2006a). *The peculiarity of self-excited oscillations in fractional systems* Acta Physica Polonica B **37**(2), 319–329.

Stanislavsky, A. A. (2006b). *Hamiltonian formalism of fractional systems* Europ. Phys. J B **49**, 93–101.

Steeb, W.-H. (2007). *Continuous symmetries, Lie algebras, differential equations and computer algebra* second edition, World Scientific Publishing, Singapore.

Strutinsky, V. M. (1967a). *Microscopic calculation of the nucleon shell effects in the deformation energy of nuclei* Ark. Fysik **36**, 629–632.

Strutinsky, V. M. (1967b). *Shell effects in nuclear masses and deformation energies* Nucl. Phys. A **95**(2), 420–442.

Strutinsky, V. M. (1968). *Shells in deformed nuclei* Nucl. Phys. **A122**, 1–33.

Stückelberg, E. C. G. (1941). *Un nuveau modèle de l'èlectron ponctuel en thèorie classique* Helv. Phys. Acta **14**, 51–80.

Szwed, J. (1986). *The square root of the Dirac equation within super symmetry* Phys. Lett. B **181**(3-4), 305–307.

Tajima, N. and Suzuki, N. (2001). *Prolate dominance of nuclear shape caused by a strong interference between the effects of spin-orbit and $l^2$ terms of the Nilsson potential* Phys. Rev. C **64**, 037301-037304.

Tarascon, J. M., Gozdz, A. S., Schmutz, C., Shokoohi, F. and Warren, P. C. (1996). *Performance of Bellcore's plastic rechargeable Lithium-ion batteries* Solid State Ionics **86-88**, 49–54.

Tarasov, V. E. and Zaslavsky, G. M. (2006). *Dynamics with low-level fractionality* Physica A **368**(2), 399–415.

Tarasov, V. E. (2008). *Fractional vector calculus and fractional Maxwell's equations* Ann. Physics **323**(11), 2756–2778.

Tatom, F. B. (1995). *The relationship between fractional calculus and fractals* Fractals **3**(1), 217–229.

Todd, S. and Latham, W. (1992). *Evolutionary art and computers* Academic Press, London.

Tofighi, A. (2003). *The intrinsic damping of the fractional oscillator* Physica A **329**, 29–34.

Treder, H. J. (1983). *On the correlations between the particles in the EPR-paradoxon* Annalen der Physik **495**, 227–230.

Uchaikin, V., Sibatov, R. and Uchaikin, D. (2009). *Memory regeneration phenomenon in dielectrics: the fractional approach* Phys. Scr. **T136**, 014002.

Vautherin, D. and Brink, D. M. (1972). *Hartree–Fock calculations with Skyrme's interaction* Phys. Rev. C **5**, 626–647.

Vinci, L. Da (1488). *De materia* Codex Atlanticus, Ambrosiana Library, Milan, Italy, f. 460v.

Wang, X. L. *et al.* Belle Collaboration (2007). *Observation of two resonant structures in $e^+e^- \to \pi^+\pi^-\psi(2S)$ via initial-state radiation at Belle* Phys. Rev. Lett. **99**(14), 142002–142007.

Weizsäcker, C. F. v. (1939). *Der zweite Hauptsatz und der Unterschied zwischen Vergangenheit und Zukunft* Annalen der Physik **428**, 275–283.

Westerlund, S. (1991). *Dead matter has memory!* Phys. Scr. **43**, 174–179.

Weyl, H. (1917). *Bemerkungen zum Begriff des Differentialquotienten gebrochener Ordnung* Vierteljahresschr. Naturforsch. Ges. Zürich, **62**, 296–302.

Whittaker, E. T. and Watson, G. N. (1965). *A course of modern analysis*, Cambridge University Press, Cambridge.

Wilson, K. G. (1974). *Confinement of quarks*, Phys. Rev. D **10**, 2445–2459.

Wiman, A. (1905). *Über den Fundamentalsatz in der Theorie der Funktionen $E_a(x)$* Acta Math. **29**, 191–201.

Wolf, G. (1980). *Selected topics on $e^+e^-$-physics* DESY 80/13.

Wolfram, S. (2010). *Wolfram Mathematica 7 documentation center* `http://reference.wolfram.com/mathematica/guide/Mathematica.html`.

Wu, G. C. and He, J. H. (2010). *Fractional calculus of variations in fractal space-time* Nonlinear Science Letters A, **1**(3), 281–287.

Yang, C. N. and Mills, R. L. (1954). *Conservation of isotopic spin and isotopic gauge invariance* Phys. Rev. **96**, 191–195.

Yonggang, K. and Xiu'e, Z. (2010). *Some comparison of two fractional oscillators* Physica B **405**, 369–373.

Young, D. A. (1978). *On the diffusion theory of phyllotaxis* J. Theor. Biology **71**(3), 421–432.

Závada, P. J. (2002). *Relativistic wave equations with fractional derivatives and pseudo-differential operators* J. Appl. Math. **2**(4), 163–197.

Zeeman, P. (1897). *The effect of magnetisation on the nature of light emitted by a substance* Nature **55**, 347–347.

Zhou, W.-X. and Sornette, D. (2002). *Generalized q-analysis of log-periodicity: Applications to critical rupture* Phys. Rev. E **66**, 046111.

Zhou, W.-X. and Sornette, D. (2003). *Non-parametric analyses of log-periodic precursors to financial crashes* International Journal of Modern Physics C **14**, 1107–1125.

Zhu, S. (2005). *The possible interpretations of Y(4260)* Phys. Lett. B **625**, 212–216.

# Index